£7.95

Materials Technology for Technicians 3

Materials Technology for Technicians 3

W. Bolton

Senior Adviser, Technician Education Council

BUTTERWORTHS
TEC
TECHNICIAN SERIES

First published 1981
Reprinted 1985

British Library Cataloguing in Publication Data

Bolton, W.

Materials technology for technicians 3.
1. Materials science
I: Title
620.1'1 TA403

ISBN 0-408-01116-5

Typeset by Tunbridge Wells Typesetting Services
Printed by the Whitefriars Press Ltd., Tonbridge & London.

Preface

This book has been written with the following aims:

1. To provide an overall view of testing methods and the interpretation of the results of such tests.
2. To describe the effects of grain size and form in metals on their properties and the factors, such as working, that affect the grains.
3. To give a basic understanding of steels and their heat treatment.
4. To describe the properties of polymers and the factors affecting those properties.
5. To explain the mechanical properties of composites in terms of their constituents.

The book covers the unit 'Materials Technology 3' of the Technician Education Council. This unit has a prerequisite of 'Materials Technology 2' which provides an introduction to the basic properties and structures of metals, both ferrous and non-ferrous, and polymers. Both the 'Materials Technology 2' unit and the unit that follows this book, i.e. 'Materials Technology 4' are covered by further books in the series.

Acknowledgements

Thanks are due to the following sources for permission to reproduce illustrations and information used in this book.

Edward Arnold Ltd, Bayer UK Limited, The British Steel Corporation, Cassell and Company Ltd, Hodder and Stoughton Limited, Arthur Lee and Sons Limited, The Macmillan Press Ltd, Sanderson Kayser Limited.

Contents

1 Mechanical testing of materials

Objectives: At the end of this chapter you should be able to: *Describe the main methods used for tensile testing, impact testing, bend tests and hardness measurements, interpreting the results of such measurements.*

THE TENSILE TEST FOR METALS

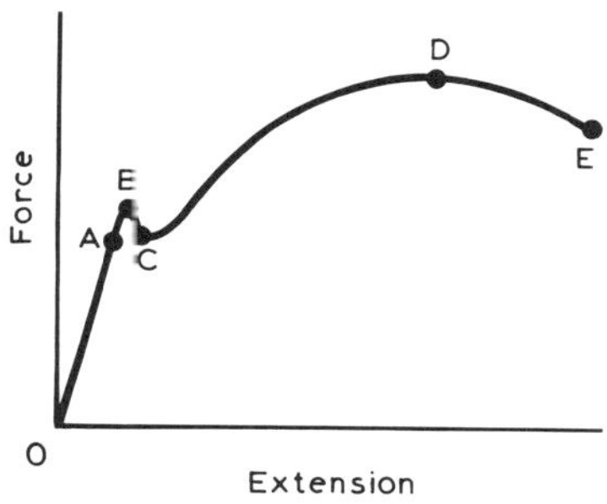

Figure 1.1 The form of a typical force-extension graph for a metal, e.g. a low carbon steel

In a tensile test, measurements are made of the force required to extend a standard size testpiece at a constant rate, the elongation of a specified gauge length of the test piece being measured by some form of extensometer. The measurements thus obtained from the test are those of force applied and resulting extension. *Figure 1.1* shows the form of a typical force-extension graph for a metal.

For the part of the force-extension graph 0A the extension is directly proportional to the force. *Hooke's law* is said to be obeyed. Point A is called the *limit of proportionality.* Up to a certain force, the material will return to its original dimensions when the force is removed, the region of the force-extension graph for which this occurs is called the *elastic region.* The maximum force for which this occurs is called the elastic limit. For many materials the elastic limit and limit of proportionality are almost identical.

Beyond the elastic limit the material will not return to its original dimensions when the force is removed, the material retaining a permanent change in dimensions. This is referred to as a *permanent set* or *plastic deformation.*

Beyond the limit of proportionality the extension is no longer proportional to the force. In some materials a situation may arise where the extension continues to increase without any increase in force; this point is known as the *yield point.* For the graph shown in *Figure 1.1,* typical of a low carbon steel, there are two distinct yield points B and C. Point B is called the *upper yield point,* point C the *lower yield point.* In data given of just yield points, with no distinction between upper and lower points, the values given are for the lower yield points.

After the yield points have been passed, an increase in force is necessary for an increase in extension; however the two are not proportional to each other. At point D the force is a maximum, after that the force decreases though the extension increases until point E is reached when the material breaks.

In order that the data obtained from one particular cross-sectional area test piece can be applied to other cross-sectional area pieces of the same material, the force information is presented as the *stress.* This stress is the force divided by the initial cross-sectional area of the test piece. The extensions are also presented as *strain,* the extension divided by the initial gauge length. The force-extension graph can thus be used to obtain the more general stress–strain graph. For stress and strain defined as above the stress–strain graph will be of the same form as the force-extension graph with just the

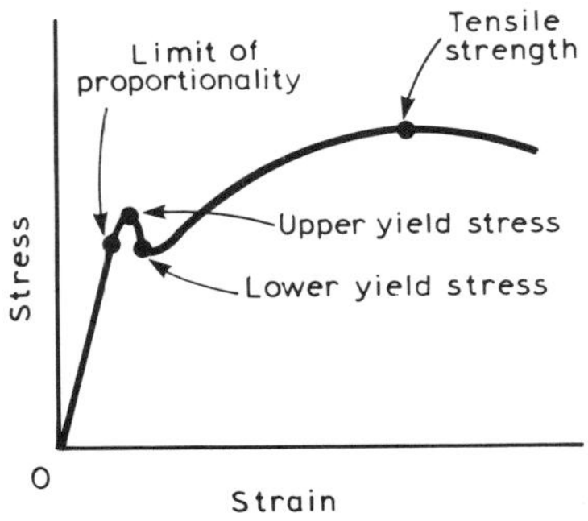

Figure 1.2 Stress–strain graph for *Figure 1.1*

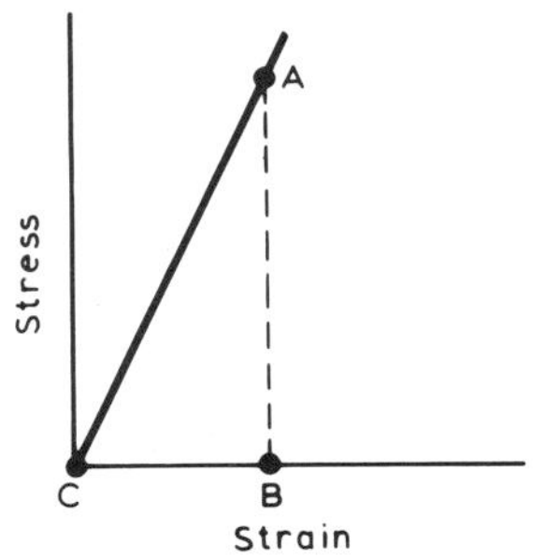

Figure 1.3 Young's modulus, or tensile modulus as it is often called, equals AB/BC

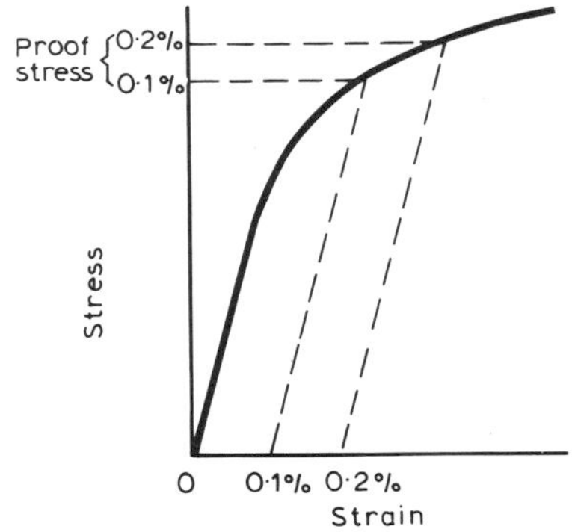

Figure 1.4 Determination of proof stress

force values divided by the initial cross-sectional area and the extension values divided by the initial gauge length (*Figure 1.2*).

For a tensile stress–strain graph the term *tensile strength* is used for the maximum force (point D on *Figure 1.1*) divided by the initial cross-sectional area of the test piece. The slope of the stress–strain graph up to the limit of proportionality, i.e. stress/strain, is called the *tensile modulus* or *Young's modulus* (*Figure 1.3*).

Many materials do not have well-defined yield points, their stress–strain graph being of the form shown in *Figure 1.4.* In such instances a *proof stress* is specified rather than a yield stress. The 0.2% proof stress is defined as that stress which results in a 0.2% offset, i.e. the stress given by a line drawn on the stress–strain graph parallel to the linear part of the graph and passing through the 0.2% strain value. Similarly the 0.1% proof stress is given by drawing a line on the stress–strain graph parallel to the linear part of the graph and passing through the 0.1% strain value.

Other data is also often obtained during a tensile test. After the test piece has broken, the pieces are fitted together and the final gauge length measured. With the initial gauge length, this enables a quantity called the *percentage elongation* to be calculated.

$$\text{Percentage elongation} = \frac{(\text{final length}) - (\text{initial length})}{\text{initial length}} \times 100$$

Similarly, another quantity called the *percentage reduction in area* can be determined from the initial cross-sectional area and the smallest cross-sectional area at fracture.

The tensile properties of metals depend on the temperature at which the data was obtained. In general the tensile modulus and the tensile strength both decrease with an increase in temperature. The percentage elongation tends to increase with an increase in temperature (see the example with *Figure 1.19*).

The data obtained from a tensile test are affected by the rate at which the test piece is strained, so in order to give standardised results, tensile tests are usually carried out at a constant strain rate. For metals tested at room temperature, the standard rate specified in BS 18 is a strain of 0.5 per minute or less.

Example

Figure 1.5 shows the stress-strain graph for a sample of mild steel. Determine (a) the limit of proportionality, (b) the upper yield point, (c) the lower yield point and (d) the tensile strength.

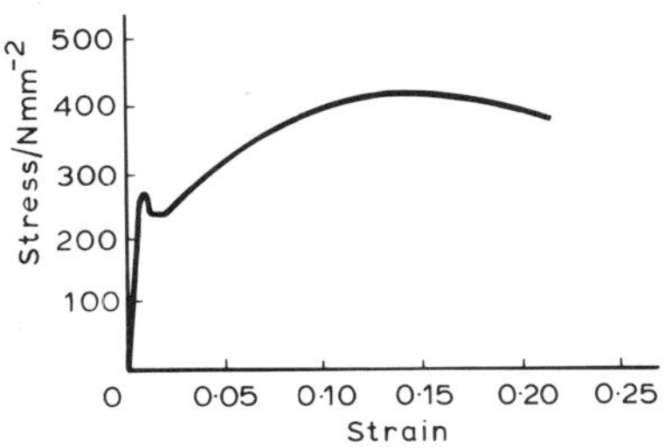

Figure 1.5 Stress-strain graph for a sample of mild steel

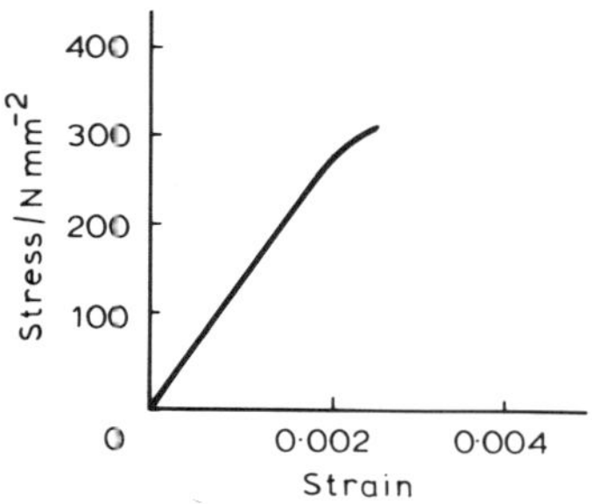

Figure 1.6 Stress–strain graph for a sample of cast iron

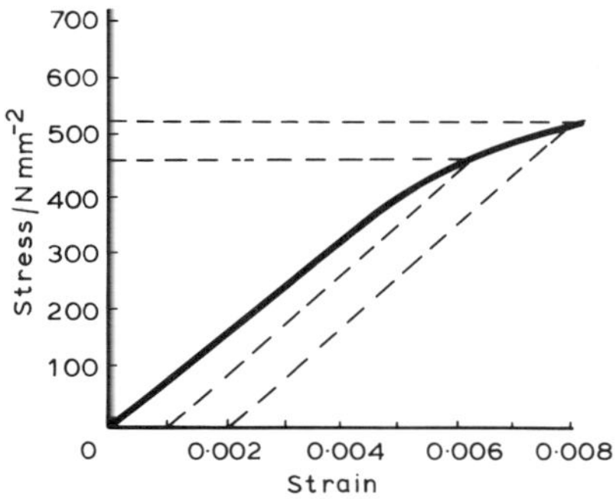

Figure 1.7 Part of the stress–strain graph for a sample of an aluminium alloy

(a) The limit of proportionality is about 240 N mm^{-2}.
(b) The upper yield point is about 280 N mm^{-2}.
(c) The lower yield point is about 240 N mm^{-2}.
(d) The tensile strength is about 400 N mm^{-2}.

Example
Figure 1.6 shows the stress–strain graph for a sample of cast iron. Determine Young's modulus for the sample.
Young's modulus is about 125 kN mm^{-2}.

Example
Figure 1.7 shows the stress–strain graph for a sample of an aluminium alloy. Determine (a) the 0.1% proof stress, (b) the 0.2% proof stress.
(a) The 0.1% proof stress is about 460 N mm^{-2}.
(b) The 0.2% proof stress is about 520 N mm^{-2}.

Example
A sample of mild steel had an initial gauge length of 69 mm. After breaking, the gauge length had become 92 mm. What is the percentage elongation?

$$\text{Percentage elongation} = \frac{(\text{final length}) - (\text{initial length})}{\text{initial length}} \times 100$$

$$= \frac{92-69}{69}$$

$$= 33\%$$

THE TENSILE TEST PIECE

In order to eliminate any variations in tensile test data due to differences in the shapes of test pieces, standard shapes are adopted. *Figure 1.8* shows the forms of two standard test pieces, one being a flat test piece and the other a round test piece. The following are the dimensions of some standard test pieces:

Flat test pieces

b/mm	L_o/mm	L_c/mm	L_t/mm	r/mm
25	100	125	300	25
12.5	50	63	200	25
6	24	30	100	12
3	12	15	50	6

Round test pieces

A/mm^2	d/mm	L_o/mm	L_c/mm	r/mm	
				Wrought material	Cast material
200	15.96	80	88	15	30
150	13.82	69	76	13	26
100	11.28	56	62	10	20
50	7.98	40	44	8	16
25	5.64	28	31	5	10
12.5	3.99	20	21	4	8

(The information in the above tables was extracted from BS 18).

For the tensile test data for the same material to give essentially the same stress–strain graph, regardless of the length of the test piece used, it is vital that the above standard dimensions be adhered to. An important feature of the dimensions is the radius given for the

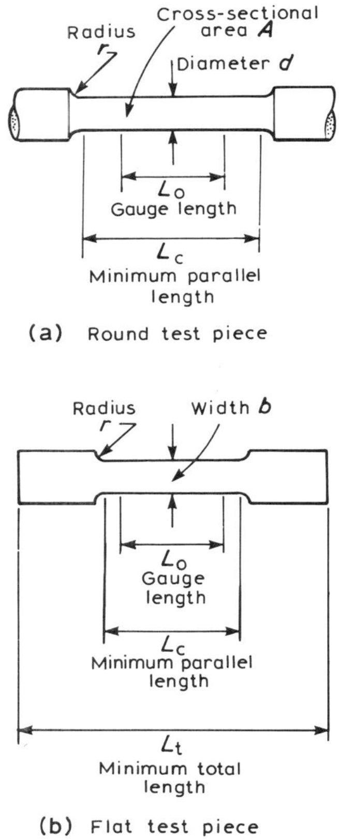

Figure 1.8 Tensile test pieces, (a) Round test piece, (b) Flat test piece

shoulders of the test pieces. Variations in the radii can affect markedly the tensile test data. Very small radii can cause localised stress concentrations which may result in the test piece failing prematurely. The surface finish of the test piece is also important for the same reason.

The round test pieces are said to be *proportional test pieces,* for which the relationship between the gauge length L_0 and the cross-sectional area of the piece A is specified in the relevant British Standard as being

$$L_0 = 5.65\sqrt{A}$$

With circular cross-sections $A = \frac{1}{4}d\pi^2$ the relationship becomes, to a reasonable approximation,

$$L_0 = 5d$$

The reason for the specification of a relationship between the gauge length and the cross-sectional area of the test piece is in order to give reproducible test results for the same test material when different size test specimens are used. When a ductile material is being extended in the plastic region of the stress–strain relationship, the cross-sectional area of the piece does not reduce uniformly but necking occurs (*Figure 1.9*). The effect of this is to cause most of the further plastic deformation to occur in the necked region where the cross-sectional area is least. The percentage elongation can thus differ markedly for different gauge lengths encompassing this necked portion of the test piece (*Figure 1.10*). Doubling the gauge length does not double the elongation because most of the elongation is in such a small part of the gauge length. The same percentage elongation is however given if

$$\frac{\text{gauge length}}{\sqrt{(\text{cross-sectional area})}} = \text{a constant}$$

In England the constant is chosen to have the fixed value of 5.65.

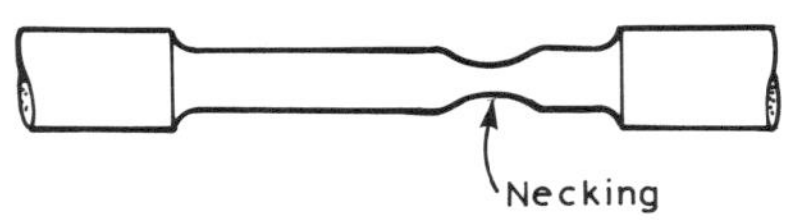

Figure 1.9 Necking of a tensile test piece

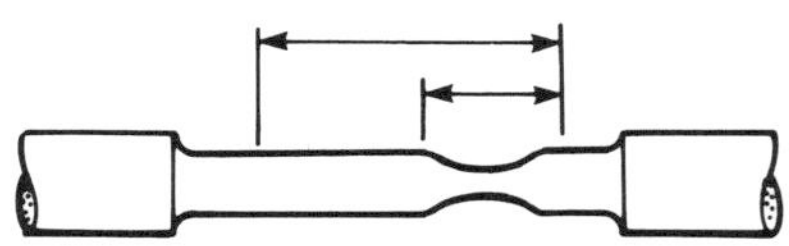

Figure 1.10 The percentage elongation for the two different gauge lengths differ considerably

The tensile test piece is usually chosen so that its properties are indicative of the properties of a component or components. This can present problems if the properties of a component are not the same in all parts of it. There can be a problem with a casting where the properties of the casting material may not be the same as that of a specially cast test piece, because of the different sizes of the two and hence the different cooling rates. Care has thus to be taken in interpreting the results of tensile test pieces.

THE TENSILE TEST FOR PLASTICS

Tensile tests can be used with plastic test pieces to obtain stress–strain data. The term tensile strength has the same meaning as with metals. The tensile modulus, i.e. Young's modulus can be determined in the same way as with metals, i.e. the slope of the stress–strain graph for

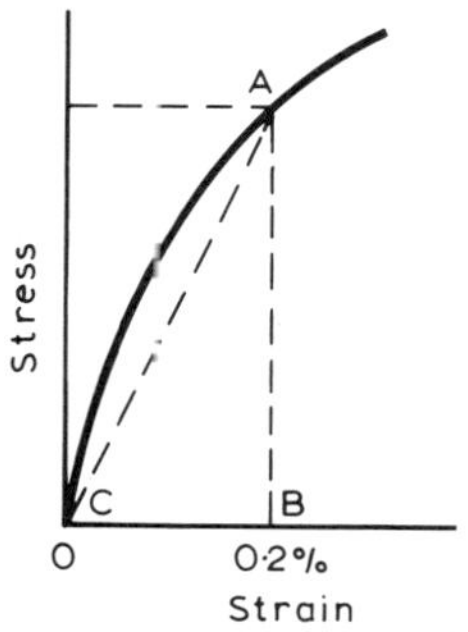

Figure 1.11 Part of the stress–strain graph for a plastic. The secant modulus is AB/BC, i.e. the slope of the line AC

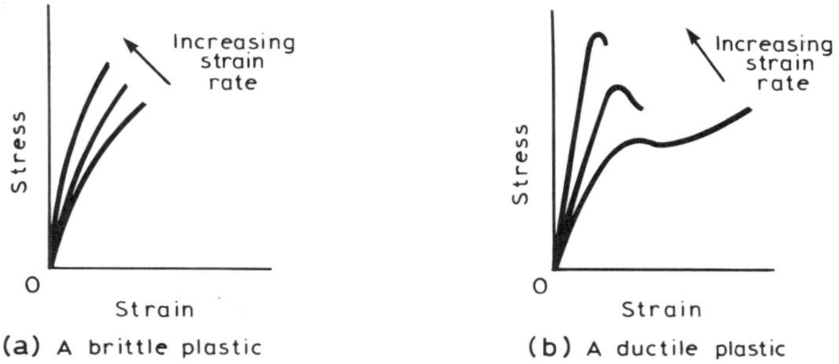

Figure 1.12 The effect of strain rate on the stress–strain graphs for plastics, (a) A brittle plastic, (b) A ductile plastic

stresses below the limit of proportionality. For many plastics there is, however, no straight-line part of the stress–strain graph and thus the tensile modulus cannot be determined in the way specified for metals. In such cases it is common practice to quote a modulus, termed the *secant modulus,* which is obtained by dividing the stress at a value of 0.2% strain by that strain (*Figure 1.11*).

The stress–strain properties of plastics are much more dependent on the rate at which the strain is applied than metals. Thus, for example, the stress–strain data may indicate a yield stress of 62 N mm^{-2} when the rate of elongation is 12.5 mm/min but 74 N mm^{-2} when it is 50 mm/min. *Figure 1.12* shows the general forms of stress–strain graphs for plastics at different strain rates. Another factor that has a considerable effect on the stress–strain properties of plastics is temperature. *Figure 1.13* shows how the tensile modulus tends to vary with temperature for plastics (this topic is discussed further in Chapter 5).

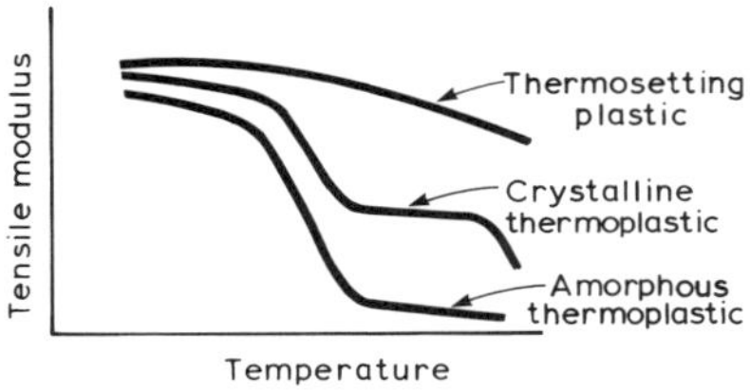

Figure 1.13 The effect of temperature on the tensile modulus of plastics

Example

Figure 1.14 is the stress–strain graph for a sample of ABS Novodur grade PK (courtesy of Bayer (UK) Ltd). Estimate (a) the tensile modulus and (b) the tensile strength.

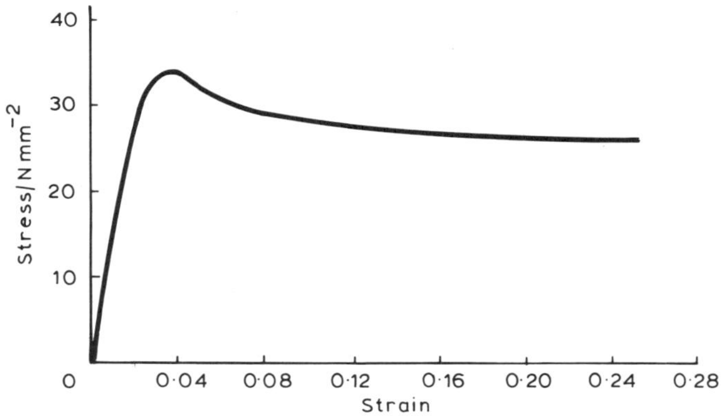

Figure 1.14 Stress–strain graph for Novodur PK. (Courtesy of Bayer UK Ltd)

(a) The tensile modulus is about 1.4 kN mm^{-2} when estimated from the graph.

(b) The tensile strength is about 34 N mm^{-2} when estimated from the graph.

INTERPRETING TENSILE TEST DATA

Figure 1.15 shows the types of stress–strain graphs produced by brittle and ductile materials. *Brittle* materials show little plastic deformation before fracture, *ductile* materials show a considerable amount of plastic deformation. If you drop a china tea cup and it breaks you can stick the pieces together again and still have the same tea cup shape. The material used for the tea cup is a brittle material and little, if any, plastic deformation took place prior to fracture. If the wing of a motor car is involved in a collision it is likely to show considerable deformation rather than a fracture. The mild steel used for the car body work is a ductile material.

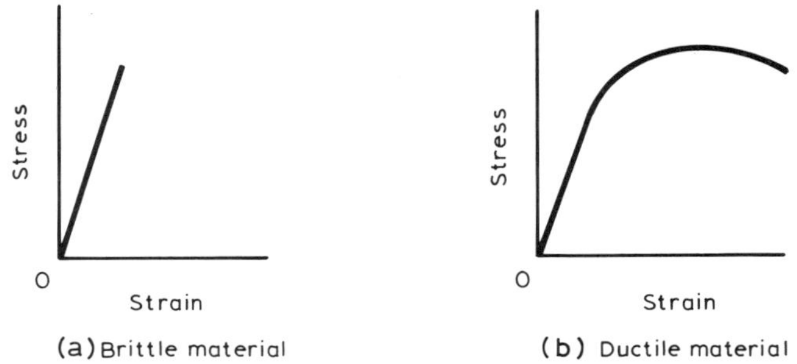

Figure 1.15 Stress–strain graphs to fracture, (a) Brittle material, (b) Ductile material

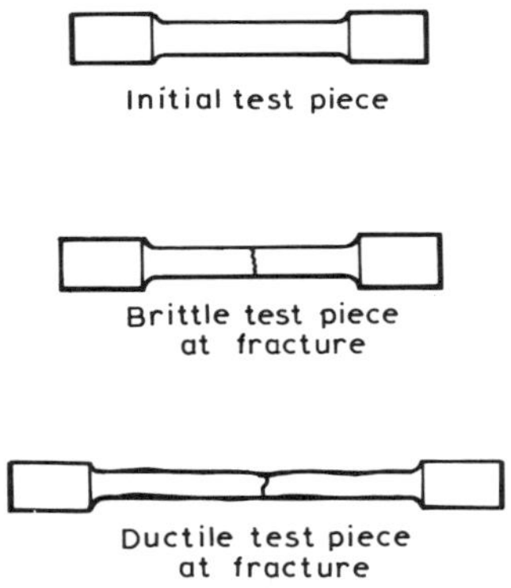

Figure 1.16

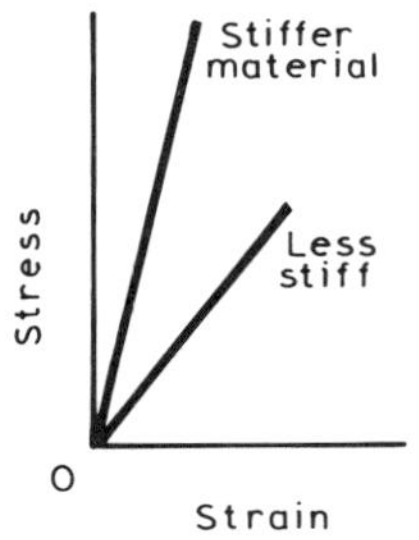

Figure 1.17 Stress–strain graphs within the limit of proportionality region

A brittle material will show only a small percentage elongation, i.e. the length of the specimen is very little different from the initial length as little plastic deformation has occurred (*Figure 1.16*). The material also shows little, if any, sign of necking. A ductile material, however, shows quite a large percentage elongation and quite significant necking.

Grey cast iron is a brittle material, it has a percentage elongation of about 0.5 to 0.7%. Mild steel is a reasonably ductile material and has a percentage elongation of the order of 30%.

Thermosetting plastics tend to behave as brittle materials. Thermoplastic materials can be either brittle or ductile depending on the temperature. Melamines are thermosetting materials and have percentage elongations of about 1% or less. High density polythene, a thermoplastic, can have percentage elongations as high as 800%.

The tensile modulus of a material can be taken as a measure of the stiffness of the material. The higher the value of the modulus the stiffer the material, i.e. the greater the force needed to produce a given strain within the limit of proportionality region (*Figure 1.17*). Mild steel has a tensile modulus of about 200 kN mm^{-2} while an aluminium alloy may have a modulus of 70 kN mm^{-2}. A strip of mild steel is thus stiffer than a corresponding strip of aluminium alloy. Plastics have relatively low tensile modulus values when compared with metals, e.g. polythene has a modulus of about 0.1 to 1.2 kN mm^{-2}.

The strength of a material is indicated by its tensile strength. An alloy steel may have a strength as high as 1500 N mm^{-2}, while an aluminium alloy may only have a strength of 200 N mm^{-2}. The steel is obviously much stronger than the aluminium alloy. Plastics have tensile strengths lower than those of metals, though their strengths can be increased by incorporating fillers in the plastic. Polythene has a tensile strength of between 4 and 38 N mm^{-2}.

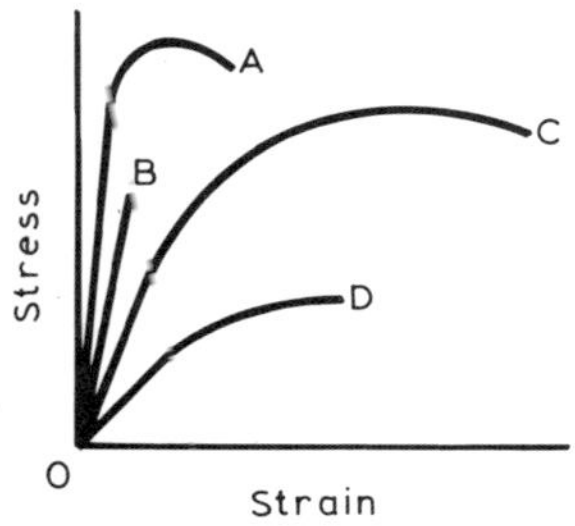

Figure 1.18

Example
Which of the materials shown in *Figure 1.18* is (a) the most ductile, (b) the most brittle, (c) the strongest, (d) the stiffest?
(a) Material C, (b) Material B, (c) Material A, (d) Material A.

Example
Which of the following materials is the most ductile?

Material	*Percentage elongation %*
80–20 brass	50
70–30 brass	70
60–40 brass	40

The most ductile material is the one with the largest percentage elongation, i.e. the 70-30 brass.

Example
Which of the following plastics is (a) the stiffest, (b) the strongest in tension?

Material	*Tensile modulus /kN mm^{-2}*	*Tensile strength /N mm^{-2}*
ABS	2.5	38
Polycarbonate	2.8	60
Polypropylene	1.3	30
PVC	3.1	50

(a) The stiffest plastic is the one with the highest tensile modulus, i.e. the PVC.
(b) The strongest plastic is the one with the highest tensile strength, i.e. the polycarbonate.

Example
Figure 1.19 shows how the tensile strength and percentage elongation varies with temperature for a carbon steel. At what temperature is the steel (a) strongest, (b) least ductile?

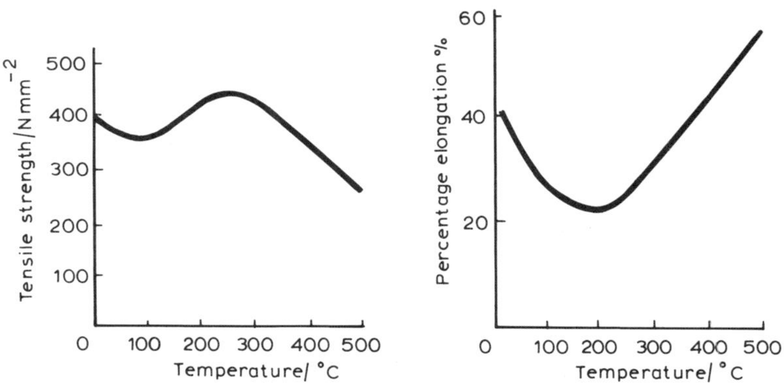

Figure 1.19 Data for a carbon steel

(a) The steel is strongest at a temperature of about 250°C, within the range for which the data is given.
(b) The steel is least ductile at a temperature of about 200°C, within the range for which the data is given.

TYPICAL TENSILE TEST RESULTS

The following table gives typical tensile strength and tensile modulus values for metals and plastics.

Material	*Tensile strength* /N mm^{-2}	*Tensile modulus* /kN mm^{-2}
Aluminium alloys	100 to 550	70
Copper alloys	200 to 1300	110
Magnesium alloys	150 to 350	45
Nickel alloys	400 to 1600	200
Titanium alloys	400 to 1600	100
Zinc alloys	200 to 350	100
Grey cast iron	150 to 400	100
Mild steel	350 to 500	200
Ferritic stainless steel	500 to 600	200
Martensitic stainless steel	450 to 1300	200
Polythene, low-density	8–16	0.2
Polythene, high-density	22–38	0.9
PVC, no plasticiser	52–58	2.7
Polystyrene	35–60	3.3
Nylon 6	70–90	2.2

IMPACT TESTS

Impact tests are designed to simulate the response of a material to a high rate of loading, and involve a test piece being struck a sudden blow. There are two main forms of test, the *Izod* and *Charpy* tests. Both tests involve the same type of measurement but differ in the form of the test pieces. Both involve a pendulum swinging down from a specified height to hit the test piece (*Figure 1.20*). The height to which the pendulum rises after striking and breaking the test piece is a measure of the energy used in the breaking. If no energy were used the pendulum would swing up to the same height as it started from. The greater the energy used in the breaking the lower the height to which the pendulum rises.

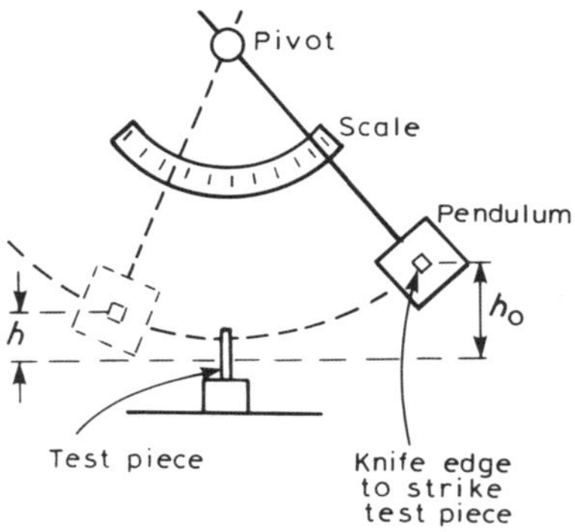

Figure 1.20 The principle of impact testing

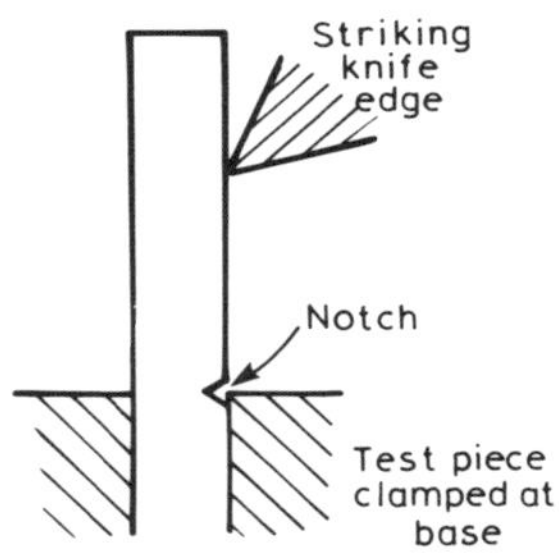

Figure 1.21 Form of the Izod test piece (elevation view)

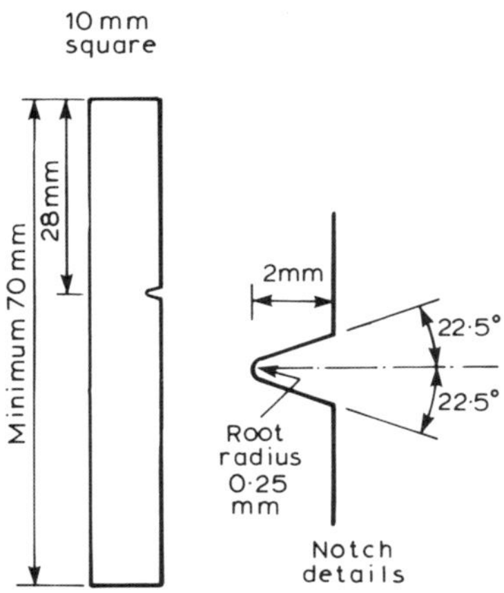

Figure 1.22 British Standard Izod test piece for a metal

With the Izod test the energy absorbed in breaking a cantilevered test piece (*Figure 1.21*) is measured. The test piece has a notch and the blow is struck on the same face of the piece as the notch is and at a fixed height above it. The test pieces used are, in the case of metals, either 10 mm square or 11.4 mm diameter if they conform to British Standards (BS 131: Part 1). *Figure 1.22* shows the details of a test piece conforming to British Standards. The British Standard test pieces for plastics (BS 2782: Part 3) are either 12.7 mm square or 12.7 mm by 6.4 to 12.7 mm depending on the thickness of the material concerned (*Figure 1.23*). With metals the pendulum strikes the test piece with a speed between 3 and 4 m/s, with plastics this speed is 2.44 m/s.

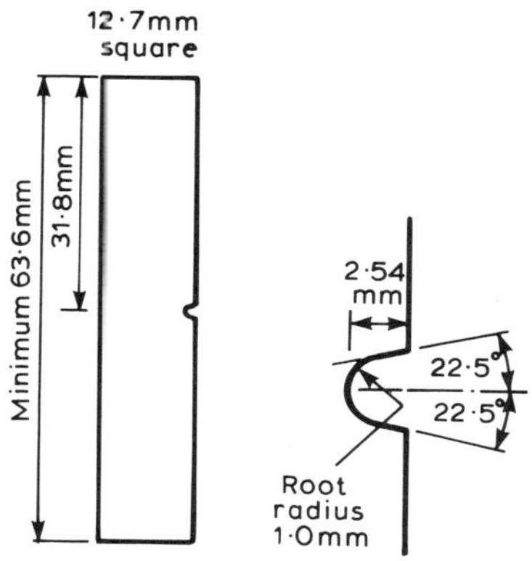

Figure 1.23 British Standard Izod test piece for a plastic

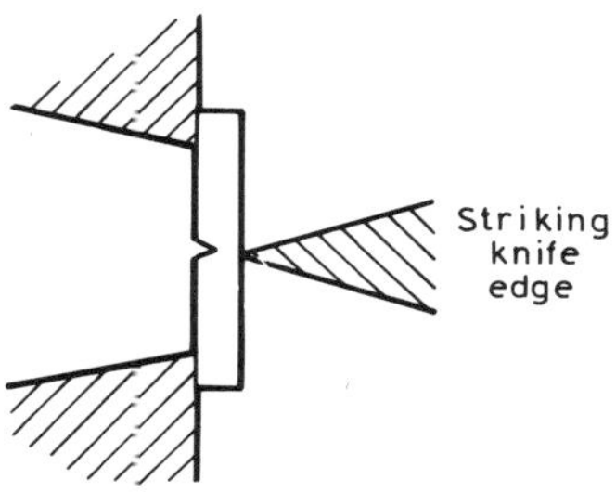

Figure 1.24 Form of the Charpy test piece (plan view)

With the Charpy test the energy absorbed in breaking a beam test piece (*Figure 1.24*) is measured. The test piece is supported at each end and is notched in the middle between the two supports. The notch is on the face directly opposite to where the pendulum strikes the test piece. For metals the British Standard test piece has a square cross-section of side 10 mm and a length of 55 mm (BS 131: Parts 2 and 3). *Figure 1.25* shows the details of a standard test piece and the three forms of notch that are possible. The results obtained with the different forms of notch cannot be compared; for comparison purposes between metals, the same type of notch should be used. The test pieces for plastics are tested either in the notched or un-notched state. The notch is produced by milling a slot across one face, the slot of width 2 mm having a radius of less than 0.2 mm at the corners of the base and the walls of the slot. A standard test piece is 120 mm long, 15 mm wide and 10 mm thick in the case of moulded plastics. Different widths and thicknesses are used with sheet plastics. With metals the pendulum strikes the test piece with a speed between 3 and 5.5 m/s, with plastics the speed is between 2.9 and 3.8 m/s.

The results of impact tests need to specify not only the type of test, i.e. Izod or Charpy, but the form of the notch used. In the case of metals the results are expressed as the amount of energy absorbed by the test piece when it breaks. The results for plastics however, are often given as absorbed energy divided by either the cross-sectional area of the un-notched test piece or the cross-sectional area behind the notch in the case of notched test pieces.

Another test that is sometimes used with plastics involves a weight falling on to a disc of the material under test. The disc, for the British Standard test (BS 2782), is 57 to 64 mm in diameter, or a square of 57 to 64 mm side resting on an annular support of 50.8 mm inner diameter and maximum external diameter of

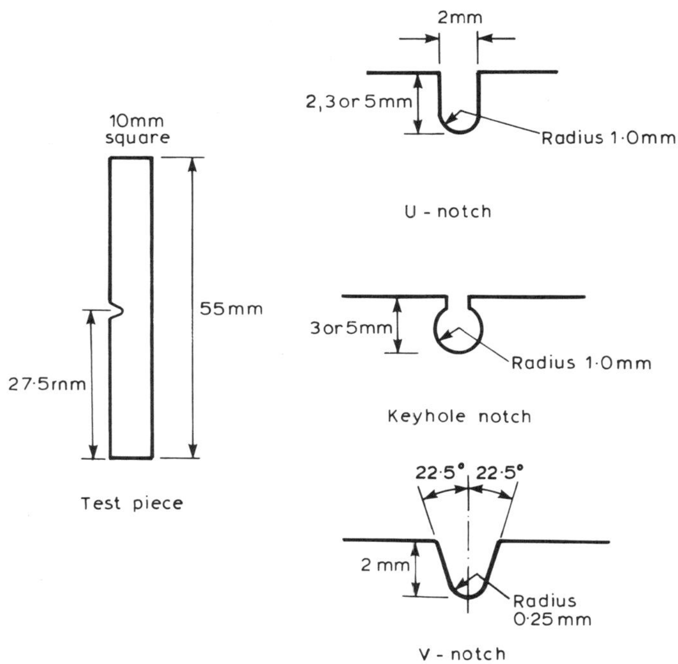

Figure 1.25 British Standard Charpy test piece for a metal

57.2 mm. The thickness of the test piece is 1.52 mm in the case of moulded plastics. The impact strength is the energy needed to fracture half of a large number of samples of the material.

Example
In an Izod test the pendulum in falling acquires a kinetic energy of 160 J before striking the test piece. If after fracturing the test piece the pendulum continues up to a height of 60% of that from which it started its motion, what is the energy used to fracture the piece?

The kinetic energy of the pendulum, after striking the test piece and breaking it, is transformed into potential energy. This potential energy is 60% of the potential energy with which the pendulum started. Thus 40% of the energy was used to break the test piece, i.e. 64 J.

INTERPRETING IMPACT TEST RESULTS

The fracture of materials can be classified roughly as either brittle or ductile fracture. With *brittle facture* there is little, or no, plastic deformation prior to fracture. With *ductile fracture* the fracture is preceded by a considerable amount of plastic deformation. Less energy is absorbed with a brittle fracture than with a ductile fracture. Thus Izod and Charpy test results can give an indication of the brittleness of materials.

The appearance of the fractured surfaces after an impact test also gives information about the type of fracture that has occurred. With a brittle fracture of metals the surfaces are crystalline in appearance, with ductile fracture the surfaces are rough and fibrous in appearance. Also with a ductile failure there is a reduction in the cross-sectional area of the test piece, but with a brittle fracture there is virtually no change in the area. With plastics a brittle failure gives fracture surfaces which are smooth and glassy or somewhat splintered, with a ductile failure the surfaces often have a whitened appearance. With plastics the change in cross-sectional area can be considerable with a ductile failure.

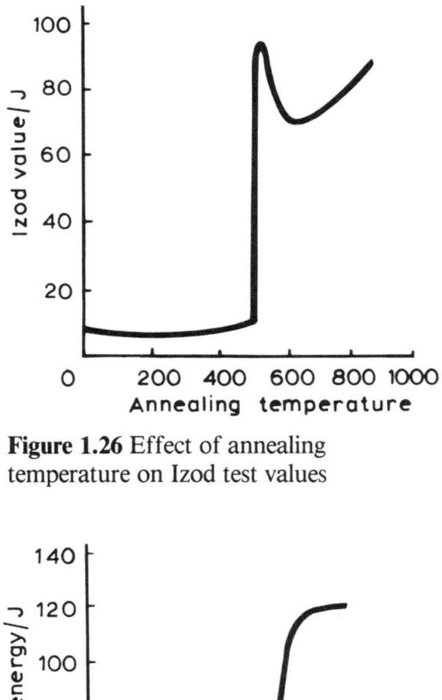

Figure 1.26 Effect of annealing temperature on Izod test values

One use of impact tests is to determine whether heat treatment has been successfully carried out. A comparatively small change in heat treatment can lead to quite noticeable changes in impact test results. The changes can be considerably more pronounced than changes in other mechanical properties, e.g. percentage elongation or tensile strength. *Figure 1.26* shows the effect on the Izod impact test results for cold worked mild steel annealed to different temperatures. The use of an impact test could then indicate whether annealing has been carried out to the correct temperature.

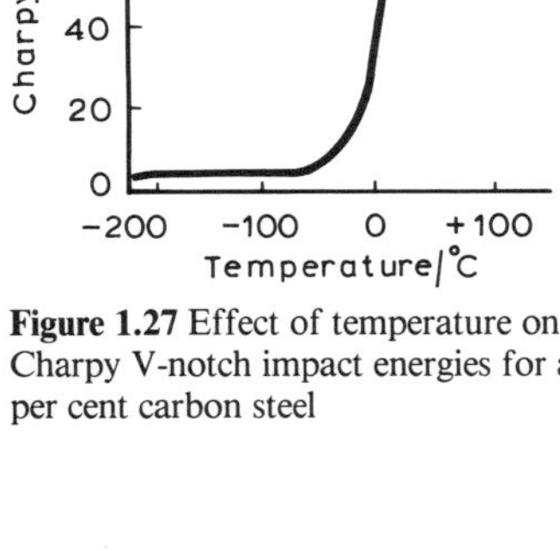

Figure 1.27 Effect of temperature on the Charpy V-notch impact energies for a 0.2 per cent carbon steel

The properties of metals change with temperature. For example, a 0.2% carbon steel undergoes a gradual transition from a ductile to a brittle material at a temperature of about room temperature (*Figure 1.27*). At about −25°C the material is a brittle material with a Charpy V-notch impact energy of only about 4 J, whereas at about 100°C it is ductile with an impact energy of about 120 J. This type of change from a ductile to a brittle material can be charted by impact test results and the behaviour of the material at the various temperatures predicted.

Example
A sample of unplasticised p.v.c. has an impact strength of 3 kJ m^{-2} at 20°C and 10 kJ m^{-2} at 40°C. At what temperature is the material most brittle?

The more brittle a material the lower the amount of energy absorbed. The material is thus more brittle at 20°C than 40°C.

Example
A structural steel gives a Charpy V-notch impact energy value of 27 J at room temperature. When the temperature drops to below 0°C this value decreases and at temperatures of −20°C it is considerably lower. Is the steel becoming more brittle or more ductile as the temperature falls?

The steel is becoming more brittle.

TYPICAL IMPACT TEST RESULTS

The following table gives typical impact test results for metals and plastics at 0°C.

Material	*Charpy V Impact strength/* J
Aluminium, commercial pure, annealed	30
Aluminium–1.5% Mn alloy, annealed	80
hard	34
Copper, oxygen free HC, annealed	70
Cartridge brass (70% Cu, 30% Zn), annealed	88
¾ hard	21
Cupronickel (70% Cu, 30% Ni), annealed	157
Magnesium–3% Al, 1% Zn alloy, annealed	8
Nickel alloy, Monel, annealed	290
Titanium–5% Al, 2.5% Sn, annealed	24
Grey cast iron	3
Malleable cast iron, Blackheart, annealed	15
Austenitic stainless steel, annealed	217
Carbon steel, 0.2% carbon, as rolled	50

Material	*Impact strength**/kJ m^{-2}
Polythene, high-density	30
Nylon 6.6	5
PVC, unplasticised	3
Polystyrene	2
ABS	25

* Notch-tip radius 0.25 mm, depth 2.75 mm

BEND TESTS

The *bend test* is a simple test of ductility. It involves bending a sample of the material through some angle and determining whether the material is unbroken and free from cracks after such a bend. *Figure 1.28* shows one way of conducting a bend test. The results of a bend test are specified in terms of the angle of bend (*Figure 1.29*).

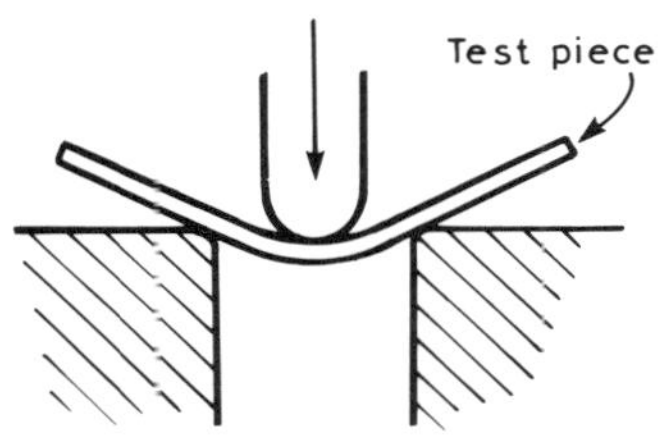

Figure 1.28 A bend test

Figure 1.29 The angle of bend

The following data is taken from a manufacturer's brochure (courtesy of Arthur Lee & Sons Ltd) and illustrates the type of information given for bend tests.

Tempers – mild steel
CS4 Hard
For maximum strength where no bending or drawing is involved. For clean and crisp shearing or blanking.

CS4 Half Hard
Strip will stand a 90° transverse bend over a radius equal to its own thickness. Useful for simple bending operations.
CS4 Quarter Hard
Strip will stand a close bend in the transverse direction and 90° longitudinal bend over a radius equal to its own thickness.
CS4 Annealed (Dead Soft)
Strip will stand being bent flat on itself in both directions, and can also be used for simple pressing and drawing.

HARDNESS MEASUREMENTS WITH METALS

The *hardness* of a material may be specified in terms of some standard test involving indentation or scratching of the surface of the material. Hardness is essentially the resistance of the surface of a material to deformation. There is no absolute scale for hardness, each hardness form of test having its own scale. Though some relationships exist between results on one scale and those on another, care has to be taken in making comparisons because the tests associated with the scales are measuring different things.

The most common form of hardness tests for metals involves standard indentors being pressed into the surface of the material concerned. Measurements associated with the indentation are then taken as a measure of the hardness of the surface. The Brinell test, the Vickers test and the Rockwell test are the main forms of such tests.

With the *Brinell test* (*Figure 1.30*) a hardened steel ball is pressed for a time of 10 to 15 s into the surface of the material by a standard force. After the load and the ball have been removed the diameter of

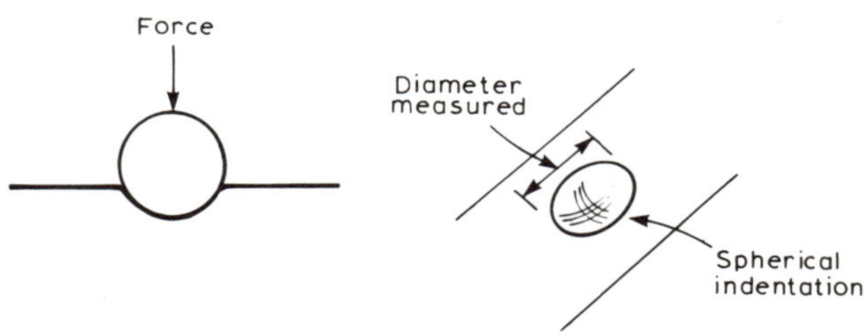

Figure 1.30 The basis of the Brinell Hardness test

the indentation is measured. The Brinell hardness number (signified by HB) is obtained by dividing the size of the force applied by the spherical surface area of the indentation. This area can be obtained, either by calculation or the use of tables, from the values of the diameter of the ball used and the diameter of the indentation.

$$\text{Brinell Hardness number} = \frac{\text{applied force}}{\text{spherical surface area of indentation}}$$

The units used for the area are mm^2 and for the force are kgf (1 kgf = 9.8 N). The British Standard for this test is BS 240.

The diameter D of the ball used and the size of the applied force F are chosen, for the British Standard, to give F/D^2 values of 1, 5, 10 or 30, the diameters of the balls being 1, 2, 5 or 10 mm. In principle, the same value of F/D^2 will give the same hardness value, regardless of the diameter of the ball used.

The Brinell test cannot be used with very soft or very hard materials. In the one case the indentation becomes equal to the diameter of the ball and in the other case there is either no or little indentation on which measurements can be based. The thickness of the material being tested should be at least ten times the depth of the indentation if the results are not to be affected by the thickness of the sample.

The *Vickers test* (*Figure 1.31*) uses a diamond indenter which is pressed for 10 to 15 s into the surface of the material under test. The result is a square-shaped impression. After the load and indenter are

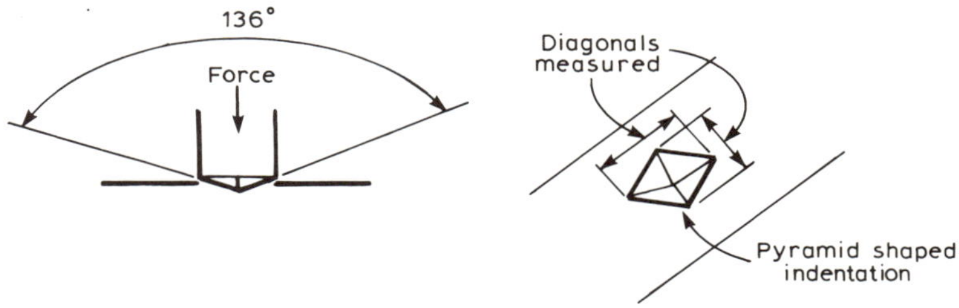

Figure 1.31 The basis of the Vickers Hardness test

removed the diagonals of the indentation are measured. The Vickers hardness number (HV) is obtained by dividing the size of the force applied by the surface area of the indentation. The surface area can be calculated, the indentation is assumed to be a right pyramid with a square base (the vertex angle of the pyramid is assumed to be the same as the vertex angle of the diamond, i.e. 136°), or obtained by using tables and the diagonal values. The relevant British Standard is BS 427.

The Vickers test has the advantage over the Brinell test of the increased accuracy that is possible in determining the diagonals of a square as opposed to the diameter of a circle. Otherwise it has the same limitations as the Brinell test.

The *Rockwell test* (*Figure 1.32*) uses either a diamond cone or a hardened steel ball as the indenter. A force is applied to press the indenter in contact with the surface. A further force is then applied

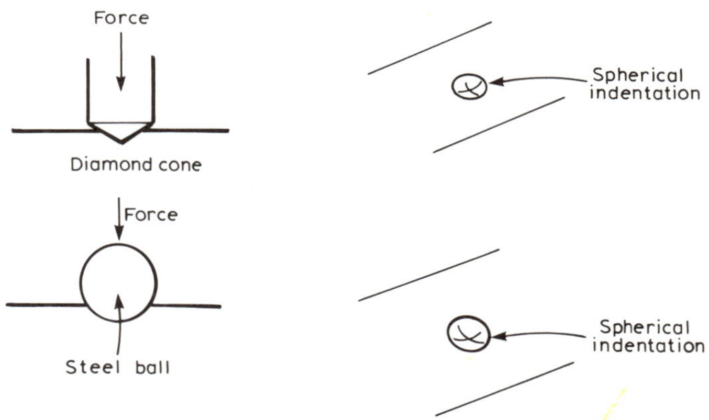

Figure 1.32 The basis of the Rockwell Hardness test

and causes an increase in depth of the indenter penetration into the material. The additional force is then removed and there is some reduction in the depth of the indenter due to the deformation of the material not being entirely plastic. The difference in the final depth of the indenter and the initial depth, before the additional force was applied, is determined. This is the permanent increase in penetration (*e*) due to the additional force.

$$\text{Rockwell Hardness number (HR)} = E - e$$

where E is a constant determined by the form of the indenter. For the diamond cone indenter, E is 100, for the steel ball, E is 130.

There are a number of Rockwell scales, the scale being determined by the indenter and the additional force used. The following table indicates the scales and the types of materials for which each are typically used.

Scale	*Indenter*	*Additional force*/kN	*Typical applications*
A	Diamond	0.59	Thin steel and shallow case-hardened steel.
B	Ball 1.588 mm dia.	0.98	Copper alloys, aluminium alloys, soft steels.
C	Diamond	1.47	Steel, hard cast irons, deep case-hardened steel.
D	Diamond	0.98	Thin steel and medium case-hardened steel.
E	Ball 3.175 mm dia.	0.98	Cast iron, aluminium alloys, magnesium alloys, bearing metals.
F	Ball 1.588 mm dia.	0.59	Annealed copper alloys, thin soft sheet metals, brass.
G	Ball 1.588 mm dia.	1.47	Malleable irons, gun metals, bronzes, copper–nickel alloys.
H	Ball 3.175 mm dia.	0.59	Aluminium, lead, zinc.
K	Ball 3.175 mm dia.	1.47	Aluminium and magnesium alloys.
L	Ball 6.350 mm dia.	0.59	Plastics.
M	Ball 6.350 mm dia.	0.98	Plastics
P	Ball 6.350 mm dia.	1.47	
R	Ball 12.70 mm dia.	0.59	Plastics
S	Ball 12.70 mm dia.	0.98	
V	Ball 12.70 mm dia.	1.47	

The relevant British Standard for Rockwell tests is B.S. 891. In any reference to the results of a Rockwell test the scale letter must be quoted. The B and C scales are probably the most commonly used for metals.

For the most commonly used indenters with the Rockwell test the size of the indentation is rather small. Localised variations of structure, composition and roughness can thus affect the results. The Rockwell test is however more suitable for workship or 'on site' use as it is less affected by surface conditions than the Brinell or Vickers tests which require flat and polished surfaces to permit accurate measurements. The test, as indicated above, cannot be used with thin sheet. A variation of the standard test, called the ***Rockwell superficial hardness test,*** can be used however with thin sheet. Smaller forces are used and the depth of the indentation is determined with a more sensitive device as much smaller indentations are used. The initial force is 29.4 N instead of 90.8 N. The following are the scales given by this test.

Scale	*Indenter*	*Additional force*/kN
15–N	Diamond	0.14
30–N	Diamond	0.29
45–N	Diamond	0.44
15–T	Ball 1.588 mm dia.	0.14
30–T	Ball 1.588 mm dia.	0.29
45–T	Ball 1.588 mm dia.	0.44

Note: numbers with the scale letter refer to the additional force values when expressed in kgf units (1 kgf = 9.8 N).

COMPARISON OF THE DIFFERENT HARDNESS SCALES

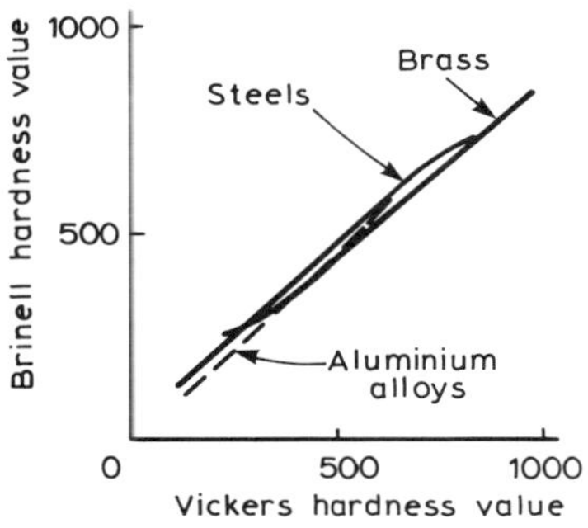

Figure 1.33 Relationships between Brinell Hardness and Vickers Hardness values

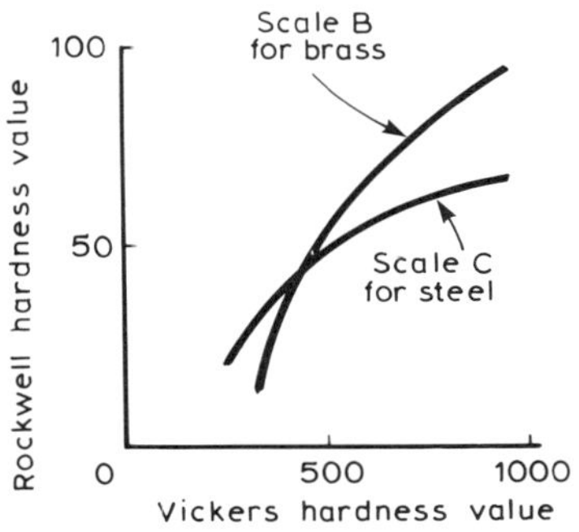

Figure 1.34 Relationships between Rockwell Hardness and Vickers Hardness values

The Brinell and Vickers tests both involve measurements of the surface areas of indentations; the form of the indenters however is different. The Rockwell tests involve measurements of the depths of penetration of the indenter. Thus the various tests are concerned with different forms of material deformation as an indication of hardness. There are no simple theoretical relationships between the various hardness scales though some approximate, experimentally derived, relationships have been obtained. Different relationships hold for different metals. *Figure 1.33* shows the types of relationship between Brinell hardness values and Vickers hardness values for steels, aluminium alloys and brass. *Figure 1.34* shows the types of relationships between Rockwell scales B and C and Vickers hardness values. The following table shows the conversions that are used between the different tests for steels.

Brinell value	*Vickers value*	*Rockwell values* B	C
112	114	66	
121	121	70	
131	137	74	
140	148	78	
153	162	82	
166	175	86	4
174	182	88	7
183	192	90	9
192	202	92	12
202	213	94	14
210	222	96	17
228	240	98	20
248	248	102	24
262	263	103	26
285	287	105	30
302	305	107	32
321	327	108	34
341	350	109	36
370	392		40
390	412		42
410	435		44
431	459		46
452	485		48
475	510		50
500	545		52

Up to a hardness value of 300 the Vickers and Brinell values are almost identical.

There is an approximate relationship between hardness values and tensile strengths. Thus for annealed steels the tensile strength in N mm^{-2} is about 3.54 times the Brinell hardness value, and for quenched and tempered steels 3.24 times the Brinell hardness value. For brass the factor is about 5.6, and for aluminium alloys about 4.2.

Example

An aluminium alloy (4.0% Cu, 0.8% Mg, 0.5% Si, 0.7% Mn) has a hardness of 45 HB when annealed and 100 HB when solution treated and precipitation hardened. What might be the tensile strengths of the alloy in these two conditions?

If the tensile strength is assumed to be the hardness value multiplied by 4.2 then the tensile strength in the annealed condition is 189 N mm^{-2} and in the heat treated condition is 420 N mm^{-2}. The measured values were 180 N mm^{-2} and 430 N mm^{-2}.

HARDNESS MEASUREMENTS WITH PLASTICS

The Brinell, Vickers and Rockwell tests can be used with plastics. The Rockwell test with its measurement of penetration depth rather than surface area of indentation is more widely used. Scale R is a commonly used scale, e.g. Nylon 6 (Durethan 30S) has a Rockwell Hardness on scale R of 120.

The British Standard test (B.S. 2782) involves an indenter, a ball of diameter 2.38 mm, being pressed against the plastic by an initial force of 0.294 N for 5 s and then an additional force of 5.25 N being applied for 30 s. The difference between the two penetration depths is measured and expressed as a *softness number.* This is just the depth expressed in units of 0.01 mm. Thus a penetration of 0.05 mm is a softness number of 5. The test is carried out at a temperature of 23 ± 1°C.

Another form of this test is the *Shore durometer.* This involves an indenter in the form of either a truncated cone with a flat end or spherically-ended cone.

THE MOH SCALE OF HARDNESS

One form of hardness test is based on assessing the resistance of a material to being scratched. With the *Moh scale* ten materials are used to establish a scale. The materials are arranged so that each one will scratch the one preceding it in the scale but not the one that succeeds it. The following is the list of the materials in this scale.

1. Talc
2. Gypsum
3. Calcspar
4. Fluorspar
5. Apatite
6. Felspar
7. Quartz
8. Topax
9. Corundum
10. Diamond

Thus Felspar will scratch apatite but not quartz. Diamond will scratch all the materials, while talc will scratch none of them.

Ten stylii of the materials in the scale are used for the test. The hardness number of a material under test is one number less than that of the substance that just scratches it. Thus, for instance, glass can just be scratched by felspar but not by apatite. The glass has thus a hardness number of 5.

TYPICAL HARDNESS VALUES

The following table gives typical hardness values for metals and plastics at room temperature.

Material	*Hardness value*
Aluminium, commercial, annealed	21 HB
hard	40 HB
Aluminium–1.25% Mn alloy, annealed	30 HB
hard	30 HB
Copper, oxygen-free HC, annealed	45 HB
hard	105 HB
Cartridge brass, 70% Cu, 30% Zn, annealed	65 HB
hard	185 HB
Magnesium alloy–6% Al, 1% Zn, 0.3% Mn, forged	65 HB
Nickel alloy, Monel, annealed	110 HB
cold worked	240 HB

Material	*Hardness value*
Titanium alloy–5% Al, 2.5% tin, annealed	360 HB
Zinc casting alloy A, as cast	83 HB
Grey cast iron	210 HB
White cast iron	400 HB
Malleable cast iron, Blackheart	130 HB
Carbon steel–0.2% carbon, normalised	150 HB
Stainless steel, austenitic (304), annealed	150 HB
cold worked	240 HB
PVC, no plasticiser	110 HRR
Polystyrene	80 HRM
ABS	70 HRM
Nylon 6	110 HRR

Figure 1.35 shows the general range of hardness values for the different types of materials when related to Vickers, Brinell, Rockwell and Moh hardness scales.

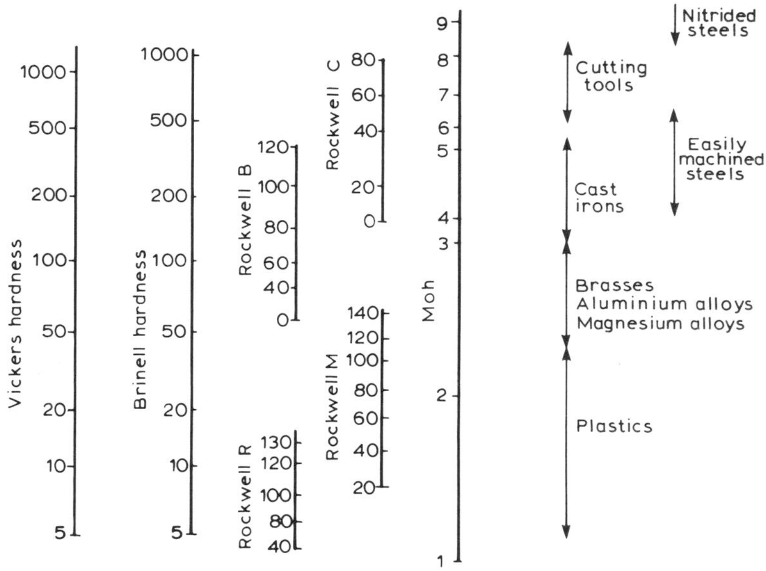

Figure 1.35 Hardness values

PROBLEMS

1. Explain the terms: tensile strength, tensile modulus, limit of proportionality, yield stress, proof stress, percentage elongation, percentage reduction in area.

2. Explain why tensile test pieces have a standard relationship between gauge length and cross-sectional area.

3. The following results were obtained from a tensile test of an aluminium alloy. The test specimen had a diameter of 11.28 mm and a gauge length of 56 mm. Determine (a) the stress-strain graph, (b) the tensile modulus, (c) the 0.1% proof stress.

Load/kN	0	2.5	5.0	7.5	10.0	12.5	15.0	17.5
Extension/10^{-2} mm	0	1.8	4.0	6.2	8.4	10.0	12.5	14.6
Load/kN		20.0	22.5	25.0	27.5	30.0	32.5	35.0
Extension/10^{-2} mm		16.3	19.0	21.2	23.5	25.7	28.1	31.5
Load/kN		37.5	38.5	39.0	39.0	(broke)		
Extension/10^{-2} mm		35.0	40.0	61.0	86			

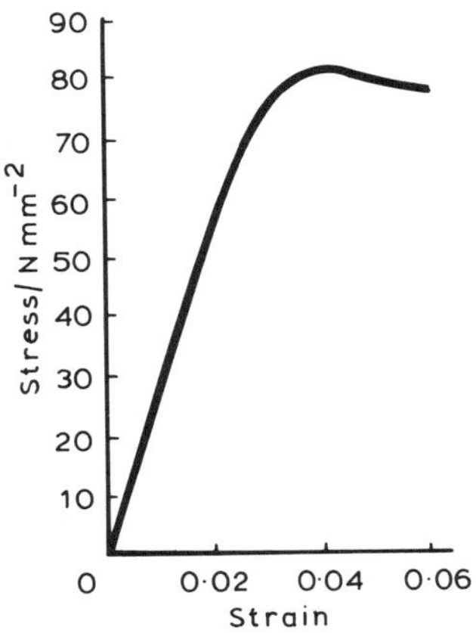

Figure 1.36 Part of the stress-strain graph for Nylon 6 (Durethan SK)

4. A flat tensile test piece of steel had a gauge length of 100.0 mm. After fracture, the gauge length was 131.1 mm. What was the percentage elongation?

5. The following data was obtained from a tensile test on a stainless steel test piece. Determine (a) the limit of proportionality stress, (b) the tensile modulus, (c) the 0.2% proof stress.

Stress/N mm^{-2}	0	90	170	255	345	495	605
Strain/10^{-4}	0	5	10	15	20	30	40
Stress/N mm^{-2}	700	760	805	845	880	895	
Strain/10^{-4}	50	60	70	80	90	100	

6. Determine from the stress–strain graph for cast iron given in *Figure 1.6*, the tensile strength and the limit of proportionality.

7. Determine from the stress–strain graph for an aluminium alloy given in *Figure 1.7,* the tensile modulus.

8. What is meant by a 'proportional test piece'?

9. *Figure 1.36* shows part of the stress–strain graph for a sample of nylon 6. Estimate (a) the tensile modulus and (b) the tensile strength for the sample.

10. What is the secant modulus of elasticity?

11. What is the effect of strain rate on the data obtained from tensile tests for plastics?

12. Sketch the form of the stress–strain graphs for (a) brittle stiff materials, (b) brittle non-stiff materials, (c) ductile stiff materials, (d) ductile non-stiff materials.

13. The effect of working an aluminium alloy (1.25% manganese) is to change the tensile strength from 110 N mm^{-2} to 180 N m^{-2} and the percentage elongation from 30% to 3%. Is the effect of the working to (a) make the material stronger, (b) make the material more ductile?

14. A sand-cast aluminium alloy (12% silicon) is found to have a percentage elongation of 5%. Would you expect this material to be brittle or ductile?

15. An annealed titanium alloy has a tensile strength of 880 N mm^{-2} and a percentage elongation of 16%. A nickel alloy, also in the annealed condition, has a tensile strength of 700 N mm^{-2} and a percentage elongation of 35%. Which alloy is (a) the stronger, (b) the more ductile in the annealed condition?

16. Cellulose acetate has a tensile modulus of 1.5 kN mm^{-2} while polythene has a tensile modulus of 0.6 kN mm^{-2}. Which of the two plastics will be the stiffer?

17. Describe the Izod or the Charpy impact test.

18. Describe the difference between the appearance of a brittle and a ductile fracture in an impact test piece.

19. Explain how impact tests can be used to determine whether a heat treatment process has been carried out successfully.

20. The following are Izod impact energies at different temperatures for samples of annealed cartridge brass (70% copper – 30% zinc). What can be deduced from the results?

Temperature/°C	+27	−78	−197
Impact energy/J	88	92	108

21. The following are Charpy V-notch impact energies for annealed titanium at different temperatures. What can be deduced from the results?

Temperature/°C	+27	−78	−196
Impact energy/J	24	19	15

22. The following are Charpy impact strengths for nylon 6.6 at different temperatures. What can be deduced from the results?

Temperature/°C	−23	−33	−43	−63
Impact strength/kJ m^{-2}	24	13	11	8

23. The impact strength of samples of nylon 6, at a temperature of 22°C, is found to be 3 kJ m^{-2} in the as moulded condition, but 25 kJ m^{-2} when the sample has gained 2.5% in weight through water absorption. What can be deduced from the results?

24. Describe a bend test and explain the significance of the results of such a test.

25. Describe the principles of the Brinell, Vickers and Rockwell hardness measurement methods.

26. How can a hardness measurement give an indication of the tensile strength of a material?

27. Outline the limitations of the Brinell hardness test.

28. With Rockwell test results, a letter A, B, C, etc. is always given with the results. What is the significance of the letter?

29. Which hardness test could be used with thin steel sheet?

30. Explain what is meant by the Moh scale of hardness.

31. A sample of brass can just be scratched by calcite but not by gypsum. What would be its Moh hardness number?

32. Specify the type of test that could be used in the following instances.
(a) A large casting is to be produced and a check is required as to whether the correct cooling rate occurs.
(b) The storekeeper has mixed up two batches of steel, one batch having been surface hardened and the other not. How could the two be distinguished?
(c) What test could be used to check whether tempering has been correctly carried out for a steel?
(d) A plastic is modified by the inclusion of glass fibres. What test can be used to determine whether this has made the plastic stiffer?
(e) What test could be used to determine whether a metal has been correctly heat treated?
(f) What test could be used to determine whether a metal is in a suitable condition for forming by bending?

2 Non-destructive testing

Objectives: At the end of this chapter you should be able to:
Describe the non-destructive testing methods commonly used and the types of information yielded by each method.

THE PURPOSE OF NON-DESTRUCTIVE TESTS

In the production of a casting, defects may occur, e.g. blow-holes, cracks, shrinkage cavities. How can the presence of such defects be established without cutting open the casting? The casting may appear externally to be perfectly sound. In the fabrication of a component defects may occur in the welds, e.g. minute cracks, lack of fusion. How can these defective welds be found? A total absence of flaws in welds is difficult and costly to achieve so some check on the quality of welds is desirable if the welded joints in components are to be of adequate quality to perform satisfactorily in service.

The above are just two examples of processes in which defects can occur. Defects may arise due to faults in a material prior to processing or during the processing. They can also occur during the service life of the manufactured component. A material defect due to, say, an incorrect heat treatment may be disclosed by taking a sample of the material and carrying out a destructive test such as a tensile test or an impact test. Thus in the case of a casting a separate test piece may be cast simultaneously. Such tests cannot disclose however the presence of internal cracks or cavities in the casting. For such purposes non-destructive testing is used. There are several methods of non-destructive testing, the commonly used ones being visual inspection, liquid penetration, magnetic and electric tests, ultrasonics and X-rays. Each of the methods has its particular characteristics and relevance to particular situations.

The use of non-destructive testing can lead to improved productivity in that defective material may be identified at an early stage in processing. The detection of faults in components can lead to better quality control, also in the case of some components greater safety, e.g. for aircraft structures. The detection of a faulty weld in the production stage could enable the fault to be rectified and a sound product produced. Non-destructive testing can thus lead to improved serviceability in that the defects which could reduce service life can be identified at an early stage.

Non-destructive testing costs money but can in itself lead to a reduction in overall costs of components. The identification of defects at an early stage enables futile processing to be avoided. Also faults may be rectified.

VISUAL INSPECTION

Visual inspection is the simplest of checks. With a welded joint it can reveal lack of penetration of a joint by a weld metal (*Figure 2.1*), the presence of large cracks that reach the surface, inadequate fusion, insufficient weld material used, poor joint shape, surface porosity.

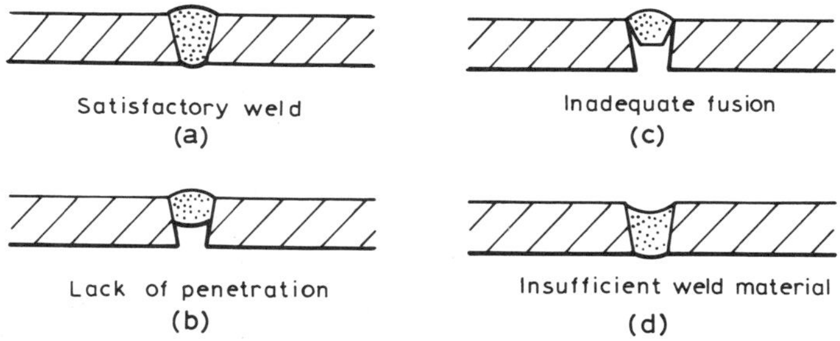

Figure 2.1 (a) Satisfactory weld, (b) Lack of penetration, (c) Inadequate fusion, (d) Insufficient weld material

LIQUID PENETRATION METHODS

While visual inspection can lead to the identification of large cracks that reach the surface, the presence of small cracks may often go unnoticed. However, cracks can be made visible by the use of suitable liquids which penetrate the cracks. The liquids are low-viscosity liquids, i.e. they flow very easily and so are able to penetrate small cracks. The component being tested has its surface covered with the liquid. After allowing sufficient time for the liquid to penetrate the cracks the surplus liquid is wiped off the surface. The surface is then 'developed'. The liquid used is generally a brightly coloured dye and the developer a fine absorbent white powder. The powder is coated over the surface as a fine layer. The powder then acts rather like a sheet of blotting paper and draws the dye up out of the cracks, so giving rise to coloured marks in the powder where the cracks are.

In some cases the liquid is applied hot. This causes the cracks to expand and so draw in more liquid, as well as making the liquid less viscous.

Another liquid penetration method involves using a fluorescent liquid. The surface under test is coated with the liquid and after a suitable interval to allow for penetration the surplus is wiped off. The surface is then viewed under ultraviolet light. In the absence of any of the liquid the surface does not glow. Any liquid leads to a glow as the liquid fluoresces. The cracks containing liquid thus show up as bright, glowing, lines against a dark background.

Ultraviolet radiation can be dangerous. While ultraviolet radiation is responsible for the sun tan desired by holiday-makers an overdose can be harmful. It is particularly dangerous for the eyes and so an ultraviolet lamp should never be viewed directly.

MAGNETIC TESTING METHODS

If you sprinkle iron filings on a bar magnet the iron filings concentrate at the ends where the magnetic lines of force emerge from the surface of the magnet (*Figure 2.2*). This is the basis of the

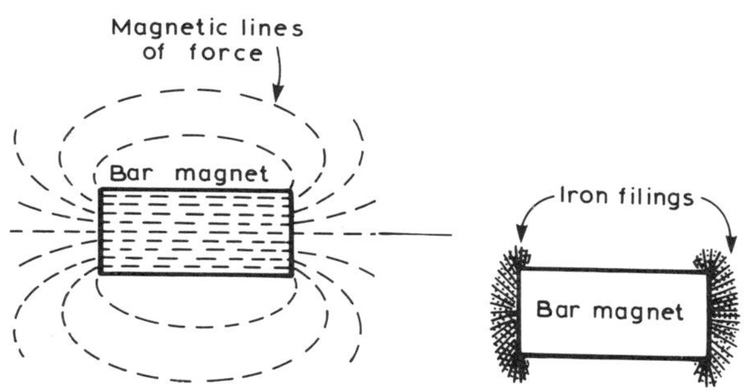

Figure 2.2

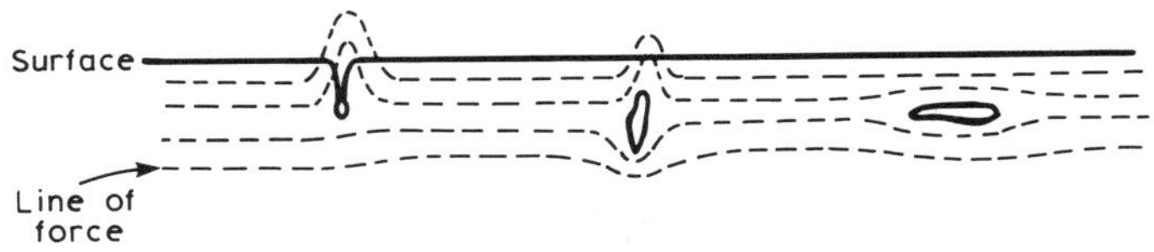

Figure 2.3 The cracks and fissures at right angles to the lines of force have the most effect on the lines of force

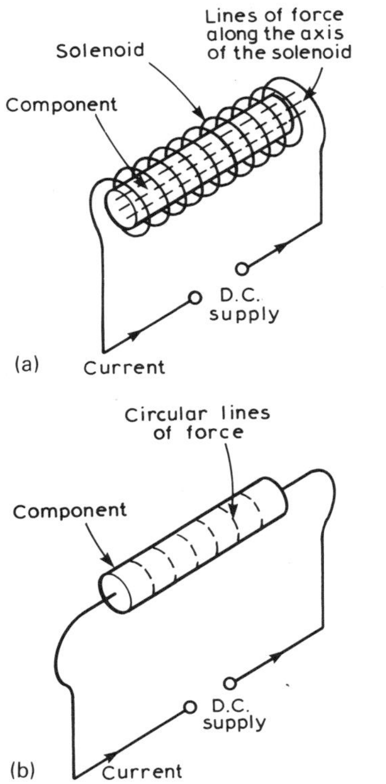

Figure 2.4 (a) Magnetisation using a current-carrying solenoid. The lines of force are parallel to the axis of the solenoid, (b) Magnetisation by passing a current through the component. The lines of force are at right-angles to the axis of the component

magnetic crack detection method. A crack in the surface, or a fissure close to the surface of a magnet, can cause the lines of force to emerge from the surface. This effect is most pronounced when the cracks or fissures are at right angles to the lines of force in the material (*Figure 2.3*).

Figure 2.4 shows two methods that can be used to magnetise components. In the first method (*Figure 2.4a*) the component is placed inside a coil of wire, i.e. a solenoid, through which a current is passed. The other method (*Figure 2.4b*) involves passing a current through the component. In the first method the lines of force are parallel to the axis of the solenoid and hence the axis of the material. This is useful where the cracks are considered to be at right angles to this axis and not parallel to the axis. With the second method the lines of force are at right angles to the axis of the component and thus of most use if the cracks are anticipated as being parallel to the axis.

The method requires a fine magnetic powder to be spread over the surface and then the concentrations of that powder looked for. In the dry method the powder, often iron oxide, is just sprinkled on to the surface. In the wet method the powder is in the form of a suspension in a light oil. This is either sprayed on to the surface or the component dipped into the suspension.

This method of crack detection is restricted to ferromagnetic materials. Iron is a ferromagnetic material. Such materials are capable of being strongly magnetised. Cracks on the surface or fissures within about 10 mm of the surface can be detected by this method.

EDDY CURRENT METHODS

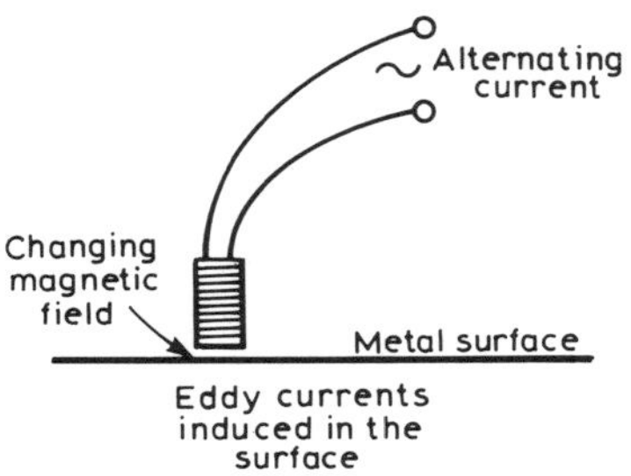

Figure 2.5 The principle of the eddy current method of non-destructive testing

If alternating current is passed through a small coil an alternating magnetic field will be produced in the vicinity of that coil. If the coil is near a piece of metal then the changing magnetic field induces currents in the metal (*Figure 2.5*). The effect is called electromagnetic induction and the currents induced in the metal are called *eddy currents.* These eddy currents also produce magnetic fields which interact with those of the coil and affect the current in the coil. If the coil is moved over the surface of a component and there are no differences in shape or composition of the metal then the current in the coil is a constant amplitude alternating current. If, however, there is a crack, the current is changed as the crack affects the eddy currents. Thus by moving the coil over the surface of the component and monitoring the alternating current in the coil, the presence of cracks and other discontinuities can be determined.

SONIC TESTING

One way to tell whether a bell is cracked is to strike it so that it sounds. An uncracked bell gives a clear note which continues for quite some time after the bell has been struck. A cracked bell however is likely to give a muffled sound which does not continue for any significant time. The presence of the crack dampens the bell vibrations. This principle can be applied to other components. Strike them a short-duration blow and listen to the sound.

ULTRASONIC METHODS

The human ear generally can detect frequencies over the range 30 to 15 000 Hz. Frequencies above 15 000 Hz are called *ultrasonic*. The frequencies used for ultrasonic non-destructive testing are generally in the region of 0.5 to 15 MHz. Such high-frequency waves have very small wavelengths. (Wave velocity v = frequency $f \times$ wavelength λ). When light waves are incident on an interface between two media, e.g. air and glass, then some of the light may be reflected and some pass on through the interface. The same situation arises with ultrasonic waves. When such waves, in passing through a block of metal, meet a fissure, i.e. a metal–air interface, then reflection can occur as well as some passing on through the interface. In practice, a very high percentage of ultrasonic waves are reflected at a metal–air interface. Thus if a pulse of ultrasonic waves is sent into a block of metal and there are no cracks or fissures, the reflected pulse will occur from the back surface of the metal. If however there is a crack there will be a reflected pulse from that point as well as from the back surface (*Figure 2.6*).

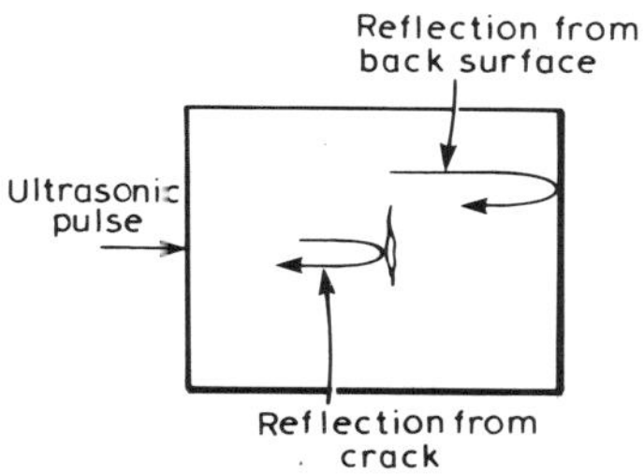

Figure 2.6 The basis of ultrasonic crack detection

By determining the time difference between the incident and the reflected pulses the presence and the position of internal cracks in a metal can be determined. The situation is essentially the same as would occur if you were to stand some distance away from a brick wall and emit a short duration sound; some time later you would hear an echo. If however there were an object intermediate between you and the wall then you would hear two echoes, one from the intermediate object and one from the wall. The presence of an echo intermediate between the initial sound and the echo from the wall tells you that there is some other object between you and the wall. The same situation occurs with ultrasonic testing for cracks.

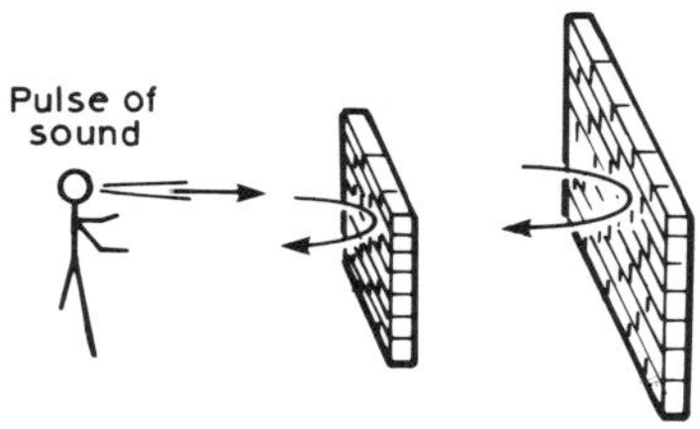

Figure 2.7 Echo sounding

Ultrasonic waves are produced by applying electrical signals to piezoelectric crystals. By using a short-duration electrical pulse a short-duration pulse of ultrasonic waves can be produced. Quartz and barium titanate are crystals that can be used. The crystal must

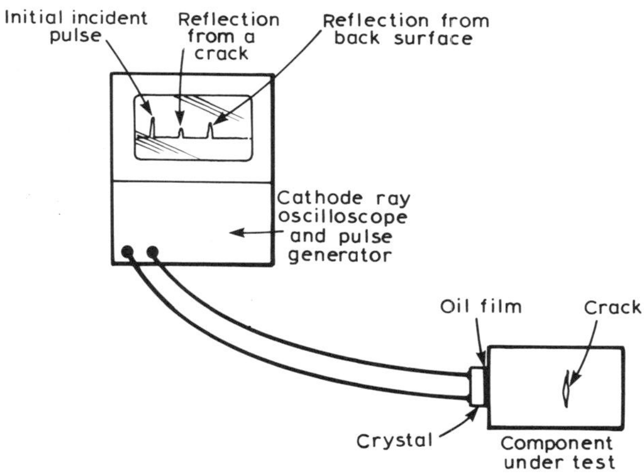

Figure 2.8 Ultrasonic crack detection

not have an air gap between it and the surface of the component being investigated, otherwise a high percentage of the pulse energy will be reflected at the outer air–metal interface; this is avoided by having a film of oil between the crystal and the metal surface. After a pulse has been emitted there is a period during which no emission of pulses occurs from the crystal. During that time the crystal is able to receive reflected pulses, which it converts into electrical signals which can then be detected. The entire procedure keeps being repeated for the train of short duration pulses emitted by the crystal. The electrical signals fed to the crystal and originating at the crystal are fed to a cathode-ray oscilloscope and displayed on the screen. Measurements of the time differences between the incident and the reflected pulses enables cracks to be located.

Ultrasonic crack detection can be used with most materials. Steel thicknesses as great as 15 m can be investigated.

RADIOGRAPHY

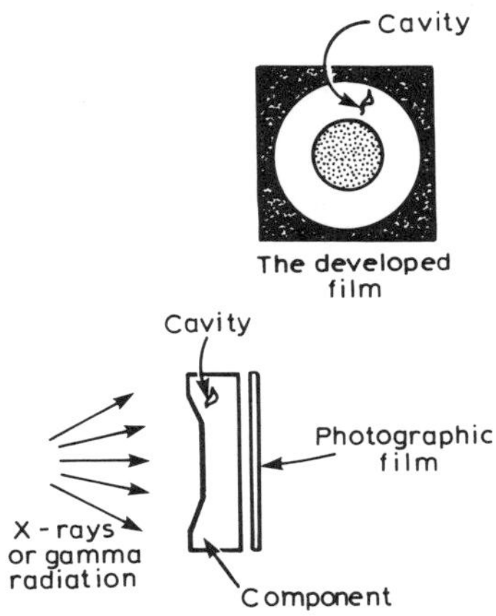

Figure 2.9 The basic principles of radiography

When X-rays or gamma radiation is passed through a material, the amount of radiation absorbed depends on the thickness and density of the material concerned. The thicker the material or the greater its density the more radiation is absorbed. Thus if there is a crack or cavity within a piece of metal the radiation passing through the metal will suffer less absorption in those regions. Similarly if there is any region of different density within the metal, e.g. a slag inclusion, there will be a difference in absorption. Basically, radiography consists of having a source of X-rays or gamma radiation on one side of a component and a piece of photographic film on the other side (*Figure 2.9*). When the film is developed the different density and different thickness parts of the component will show up as different degrees of blackening of the film.

The smallest defect that can be detected by this method is of the order of 2% of the section thickness. Thus for a component section thickness of 10 mm a defect 0.2 mm or larger can be detected. The thickness of the section through which the radiation can be passed depends on the energy of the radiation used. For X-rays the energy is determined by the voltage applied to the X-ray tube. The following table shows how the thickness of steel required to reduce the amount of radiation by a half depends on the tube voltage.

Tube voltage/kV	*Half value thickness/mm*
150	4
250	6

Thus for X-rays from a tube operating at 250 kV a thickness of steel of 6 mm will reduce the amount of radiation by a half; an extra 6 mm will reduce the radiation by a further half, i.e. down to a quarter of that incident on the steel. Thus 12 mm of steel reduces the amount of radiation to a quarter and with 18 mm of steel the radiation is reduced to one eighth.

With gamma radiation the source is a piece of radioactive material. Unlike X-ray tubes the radioactive material cannot be 'switched off' and so is constantly emitting radiation even when it is not being used for radiography. Iridium 192 and cobalt 60 are two radioactive sources that are commonly used. The following are the half-value thicknesses of steel for these sources.

Radioactive source	*Half value thickness/mm*
Iridium 192	11
Cobalt 60	22

The radioactive sources do not, unlike an X-ray tube, give an output which is constant, the strength of the source decays with time. The useful life of iridium 192 is about 40 days, that of cobalt 60 about 5.3 years. Radioactive sources do however have an advantage over X-rays tubes of not requiring voltage supplies and being very compact and so able to operate in restricted spaces.

In choosing the voltage to be used with the X-ray tube, or the type of radioactive source, the thickness and type of material to be investigated must be considered. There must be a significant reduction in the strength of the radiation in passing through the material but not so great that virtually no radiation emerges from the material. This generally means a thickness of material varying from about one half-value thickness up to about eight or nine half-value thicknesses. Thus with 150 kV X-rays the steel can be from about 4 mm to 32 to 36 mm thick. With Cobalt 60 the steel can vary from about 22 mm to 196 or 208 mm. Other materials have different half value thicknesses for these radiations.

Stringent safety precautions have to be adopted when either X-rays or gamma radiation is used. The operator of the X-ray tube or gamma source must be fully protected from the radiation, this means that there must be sufficient material between operator and radiation source for the strength of the radiation to be reduced to a very small amount on passing through the material. Because lead has a very low half-value thickness, i.e. this means a small thickness of lead absorbs a lot of radiation, lead shielding is widely used. Because the gamma source does not cease radiating when not in use it must be stored carefully so that there is no hazard to anybody.

PROBLEMS

1. Explain why non-destructive testing is widely used in industry.

2. Outline the form of the following tests and the types of defects for which they are most useful.
(a) Liquid penetrants.
(b) Ultrasonics.
(c) Magnetic powders.
(d) Radiography.

3. Which type of test or tests would be most suitable for detecting the following faults.
(a) Entrapped slag in a casting.
(b) Inadequate penetration with a weld.
(c) Surface cracks in a casting.
(d) Inclusions in rolled steel plate.

3 The grain structure of metals

Objectives: At the end of this chapter you should be able to:
Describe the effect of hot and cold work on a metal structure.
Describe the effect of heat on cold worked metals.
Describe the effect of grain size and shape on the mechanical properties of metals.
Predict the type of grain structure, and hence mechanical properties, that will occur for given work and temperature conditions.

PLASTIC DEFORMATION OF METALS

When a material is stretched beyond its elastic limit, plastic deformation can occur (see page 1 and *Figure 1.1*). With plastic deformation a material does not return to its original dimensions when the applied stress is removed but is permanently deformed. An important group of forming processes used with metals involves shaping the metals by plastic deformation. These processes include rolling, drawing, pressing, spinning, forging and extrusion.

What happens to a metal when plastic deformation occurs? Metals are crystalline. This means that the metal atoms are arranged in a regular manner within the crystal. This regular array of atoms enables us to consider the atoms in a crystal as lying in a number of planes (*Figure 3.1*). The different planes have different densities of

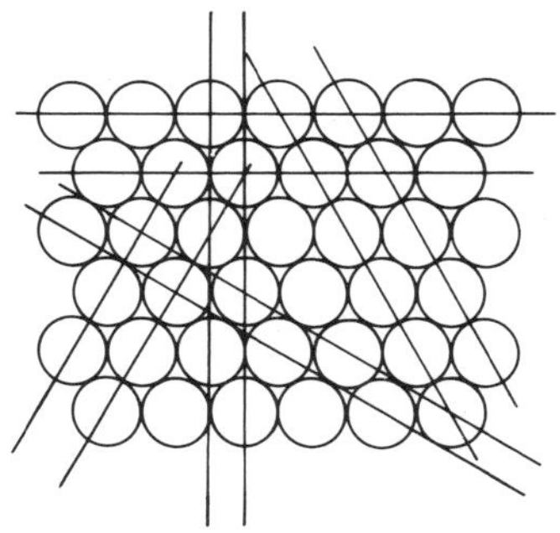

Figure 3.1 Some of the possible planes in a regularly packed array of atoms

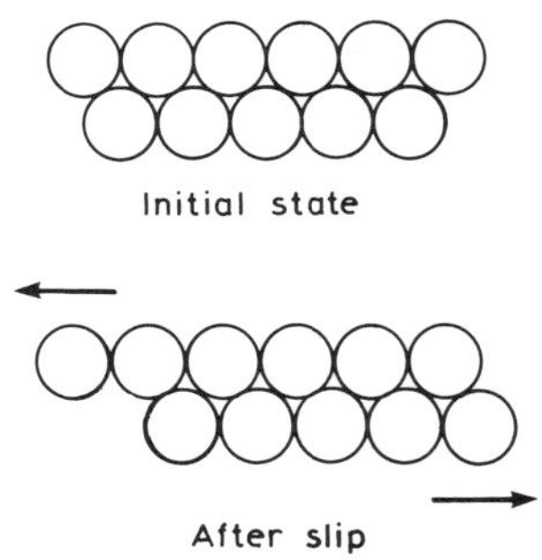

Figure 3.2 Slip of atom planes

atoms and different spacings between parallel, like, planes. When plastic deformation occurs we can consider planes of atoms to be sliding past each other (*Figure 3.2*). The plane along which this movement occurs is called a *slip plane.* This slip of planes tends to occur between those parallel planes with the highest atomic density and the greatest separation (*Figure 3.3*).

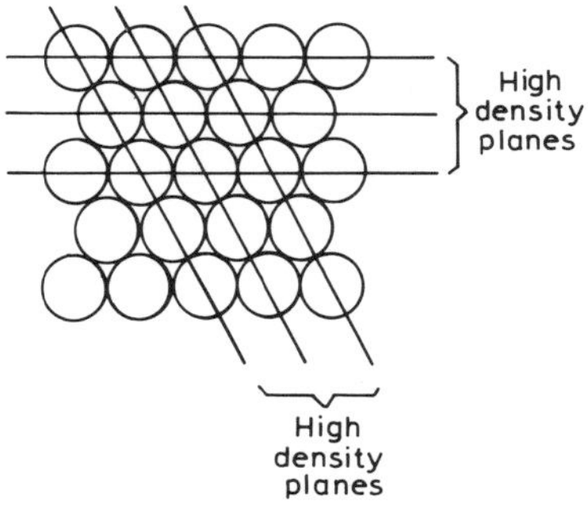

Figure 3.3

Metals are composed of many crystals. A crystal within a metal is just a region of orderly packed atoms. Such a region is generally referred to as a *grain.* The surfaces that divide the different regions of orderly packed atoms are termed *grain boundaries.* When plastic deformation occurs in a metal, movement occurs along slip planes and the result is rather like *Figure 3.4.* Slip occurs only in those planes which are at suitable angles to the applied stress. The result is

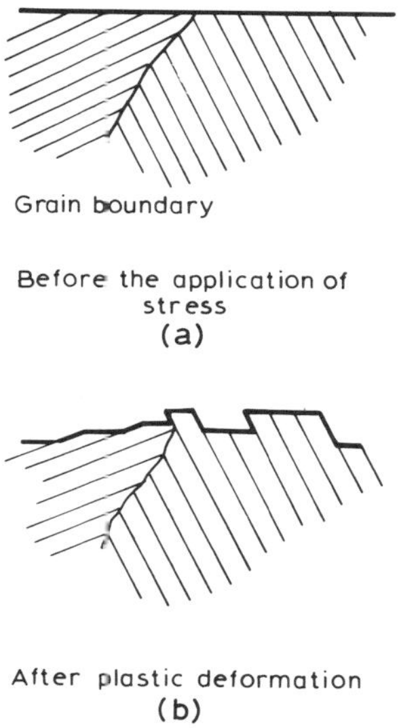

Figure 3.4 (a) Before the application of stress, (b) After plastic deformation

that the surface of the metal shows a series of steps due to the different movements of the various planes of atoms. These can be seen under a microscope (*Figure 3.5*). As will be seen from the figure the slip lines do not cross over from one grain to another; the grain boundaries restrict the slip to within a grain. Thus the bigger the grains the more slippage that can occur; this would show itself as a greater plastic deformation. A fine grain structure should therefore have less slippage and so show less plastic deformation, i.e. be less ductile. A brittle material is thus one in which each little slip process is confined to a short run in the metal and not allowed to spread, a ductile material is one in which the slip process is not confined to a short run and does spread over a large part of the metal.

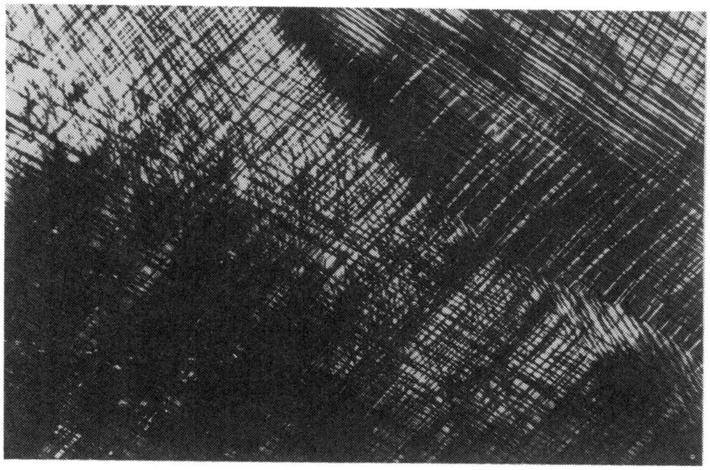

Figure 3.5 Slip steps in polycrystalline aluminium. (Courtesy of the Open University)

COLD WORKING

Suppose you were to take a carbon steel test piece and perform a tensile test on it. You may, for instance, find that the material showed a yield stress of 430 N mm^{-2}. If the test is continued beyond this point but the stress released before the tensile strength is reached a permanent deformation of the test piece will be found to have occurred. *Figure 3.6* illustrates this sequence of events. Suppose you now repeat the test. This time the yield stress is not 430 N mm^{-2} but 550 N mm^{-2}. The material has a much higher yield stress (*Figure 3.7*). This phenomena is called *cold working*.

It is not only the yield stress that changes during cold working and it is not only carbon steel that shows such changes. The term cold

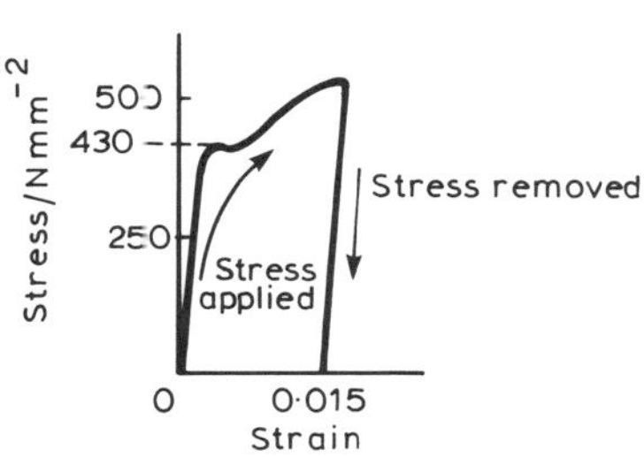

Figure 3.6

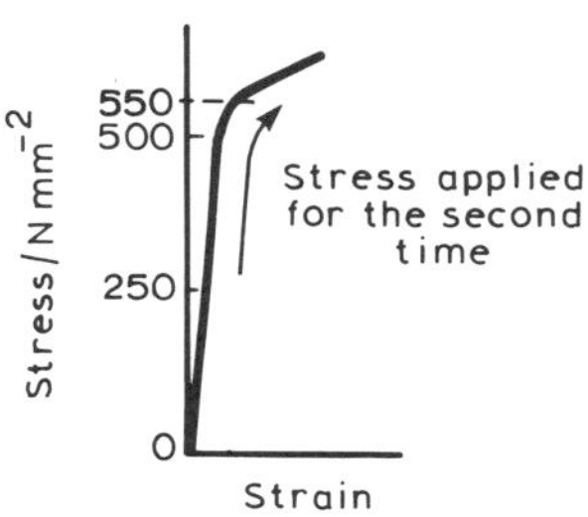

Figure 3.7

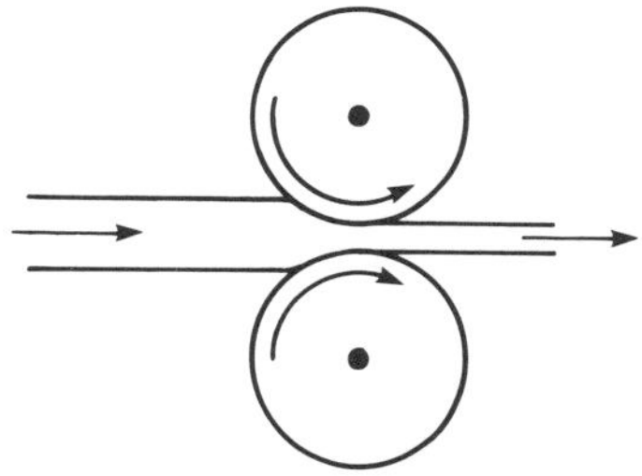

Figure 3.8 Sheet passing through rollers and being plastically deformed

working is applied to any process which results in plastic deformation at a temperature which does not alter the structural changes produced by the working. The following table shows some of the changes that take place when a sheet of annealed aluminium is rolled and its thickness reduced (*Figure 3.8*).

Reduction in sheet thickness %	*Tensile strength* /N mm^{-2}	*Elongation* %	*Hardness* HV
0	92	40	20
15	107	15	28
30	125	8	33
40	140	5	38
60	155	3	43

(Based on a table in John, V. B., *Introduction to Engineering Materials* — MacMillan)

As the amount of plastic deformation is increased so the tensile strength increases, the hardness increases and the elongation decreases. The material is becoming harder as a result of the cold working, hence the term sometimes applied to cold working of *work hardening.* The more the material is worked the harder it becomes. Also, as the percentage elongation results above indicate, the more a material is worked the more brittle it becomes. A stage can however be reached when the strength and hardness are a maximum and the elongation a minimum and further plastic deformation is not possible, the material is too brittle. With the rolled aluminium sheet referred to in the above table, this condition has been reached with about a 60% reduction in sheet thickness. The material is then said to be *fully work hardened.*

Figure 3.9 shows how, for a number of materials, the hardness depends on the amount of cold working. The percentage reduction in thickness of a sheet or the percentage reduction in cross-sectional area of other forms of material is taken as a measure of the amount of working. As will be seen from the graph, as the amount of cold working increases the hardness increases until at some value of cold working a maximum hardness is reached. The material is then fully work hardened.

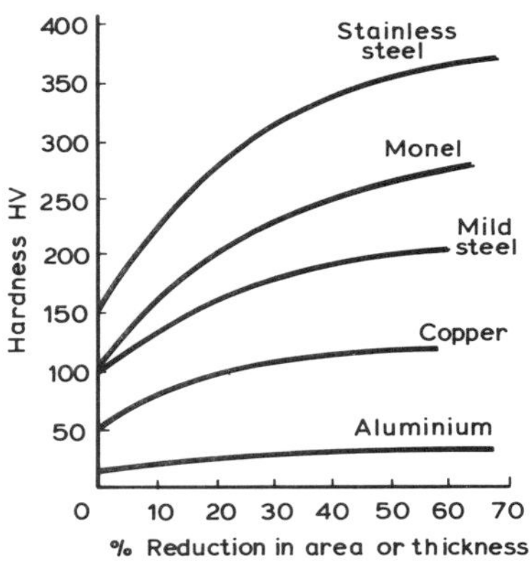

Figure 3.9 The effect of cold working on hardness

Example

Using *Figure 3.9,* what is the approximate percentage reduction in thickness of a sheet of mild steel that is possible before it becomes fully work hardened?

The mild steel would appear to have reached its maximum hardness with a percentage reduction in thickness of about 50 to 60%.

THE STRUCTURE OF COLD WORKED METALS

When stress is applied to a metal grain, deformation starts along the slip planes most suitably orientated. The effect of this is to cause the grains to become elongated and distorted. *Figure 3.10* shows the results of heavy rolling of a tin bronze ingot (4% tin). The grains have become elongated into fibre-like structures, which has the effect of giving the material different mechanical properties in different directions, a greater strength along the grain than at right angles to the grain. This effect can be used to advantage by the designer.

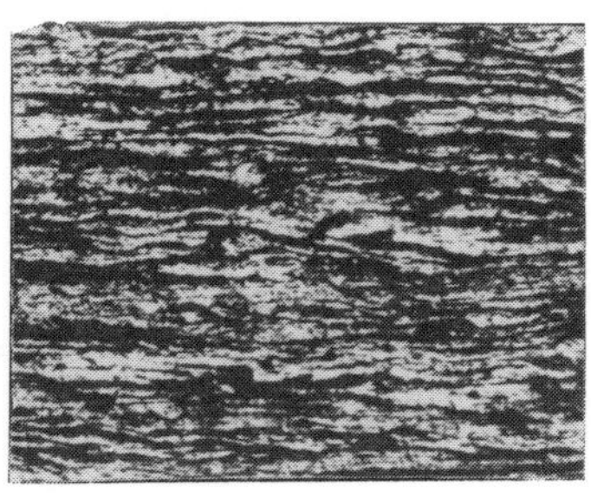

Figure 3.10 The grains in a heavily rolled tin bronze ingot. (From Rollason, E.C. *Metallurgy for Engineers,* Edward Arnold)

THE EFFECT OF HEAT ON COLD-WORKED METALS

Cold-worked metals generally have deformed grains and have often become rather brittle due to the working. In this process of deforming the grains, internal stresses build up.

When a cold-worked metal is heated to temperatures up to about 0.3 T_m, where T_m is the melting point of the metal concerned on the Kelvin scale of temperature, then the internal stresses start to become relieved. There are no changes in grain structure during this but just some slight rearrangement of atoms in order that the stresses become relieved. This process is known as *recovery*. Copper has a melting point of 1083°C, or 1356 K. Hence stress relief with copper requires heating to above about 407 K, i.e. 134°C.

If the heating is continued to a temperature of about 0.3 to 0.5 T_m there is a very large change in hardness. The strength and also the structure of the metal change. The following table shows how the hardness of copper changes, the copper having been subject to a 30% cold working.

Temperature /°C	/K		*Hardness* HV
Initially			86
150	423	(0.3 T_m)	85
200	473		80
250	523	(0.4 T_m)	74
300	573		61
350	623	(0.5 T_m)	46
450	723		24
600	873	(0.6 T_m)	15

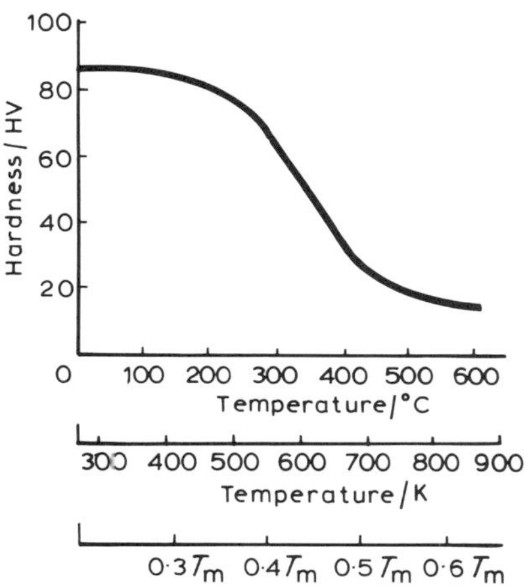

Figure 3.11 The effect of heat treatment on cold worked copper

Figure 3.11 shows the above results graphically. Between 0.3 T_m and 0.5 T_m there is a very large change in hardness. The strength also decreases while the elongation increases. What is happening is that the metal is recrystallising.

With *recrystallisation* crystals begin to grow from nuclei in the most heavily deformed parts of the metal. The temperature at which recrystallisation just starts is called the *recrystallisation temperature.* This is, for pure metals, about 0.3 to 0.5 T_m.

Material	*Melting point* /°C	/K	*Recrystallisation temperature* /°C	/K	
Aluminium	660	933	150	423	0.5 T_m
Copper	1083	1356	200	473	0.3 T_m
Iron	1535	1808	450	723	0.4 T_m
Nickel	1452	1725	620	893	0.5 T_m

As the temperature is increased from the recrystallisation temperature so the crystals grow until they have completely replaced the original distorted cold worked structure. *Figure 3.12* illustrates this sequence and its relationship to the changes in physical properties.

The sequence of events that occur when a cold worked metal is heated can be broken down into three phases:

(1) Recovery. The only significant change during this phase is the relief of internal stresses.
(2) Recrystallisation. The hardness, tensile strength and percentage elongation all change noticeably during this phase.
(3) Grain growth. The hardness, tensile strength and percentage elongation change little during this phase. The only change is that the grains grow and the material becomes large-grained.

During the *grain-growth* phase the newly formed grains grow by

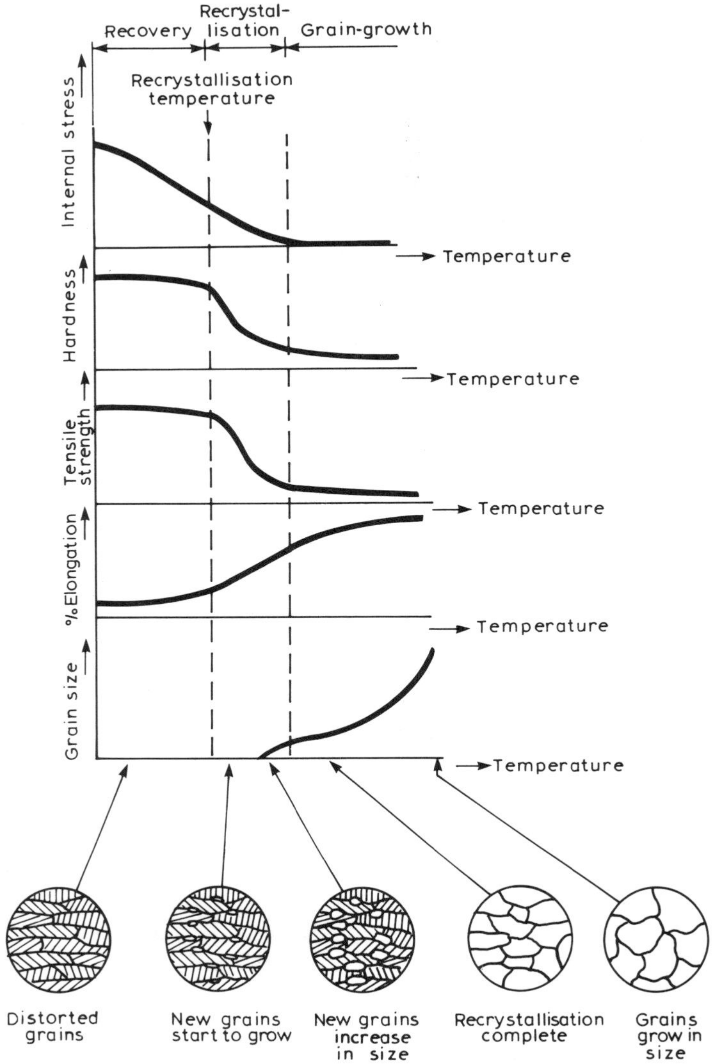

Figure 3.12 The effect of an increase in temperature on cold worked materials

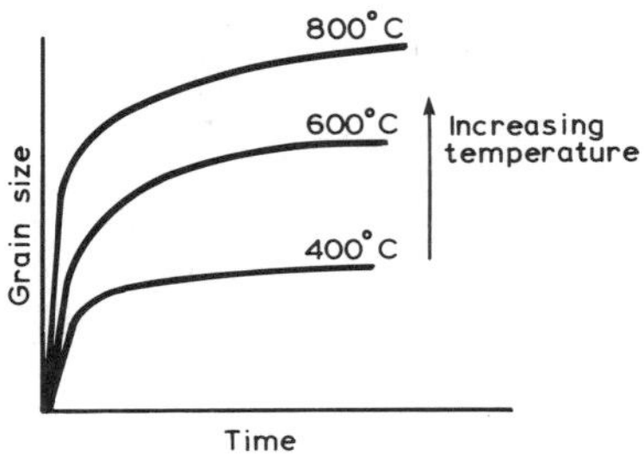

Figure 3.13 The effect of time and temperature on grain growth

absorbing other neighbouring grains. The amount of grain growth depends on the temperature and the time for which the material is at that temperature (*Figure 3.13*).

The term *annealing* is used for the heating process used to change the properties of a material. Thus, in the case of aluminium that has been cold worked and has become too brittle to work further, heating to above the recrystallisation temperature of 150°C enables new grains to grow and the material to become more ductile. The aluminium can then be worked further. This sequence of events, cold working followed by annealing and then further cold working, is used in many manufacturing processes.

Example
What is the minimum temperature at which recrystallisation occurs for cold worked copper?

The table on page 29 indicates the recrystallisation temperature as being 200°C for copper.

Example
A manufacturer of copper sheet receives the copper as much thicker plate, with already some amount of cold working. He proposes to produce the sheet by cold rolling in a number of stages, the stages being separated by annealing. Why is the sheet production in a number of stages?

If copper is cold-worked to about a 60% reduction in cross-sectional area it becomes brittle and tends to break with further working. Also at reductions approaching the 60% the material is becoming fairly hard and difficult to roll. By following rolling by annealing the material is made more ductile again and rolling can continue without difficulty.

FACTORS AFFECTING RECRYSTALLISATION

(1) A minimum amount of deformation is necessary before recrystallisation can occur. The permanent deformation necessary depends on the metal concerned.
(2) The greater the amount of cold work the lower the crystallisation temperature for a particular metal.
(3) Alloying increases the recrystallisation temperature.
(4) No recrystallisation takes place below the recrystallisation temperature. The higher the temperature above the recrystallisation temperature the shorter the time needed at that temperature for a given crystal condition to be attained.
(5) The resulting grain size depends on the annealing temperature. The higher the temperature the larger the grain size.
(6) The amount of cold work prior to the annealing affects the size of the grains. *Figure 3.14* shows the effect on the grain size of annealing

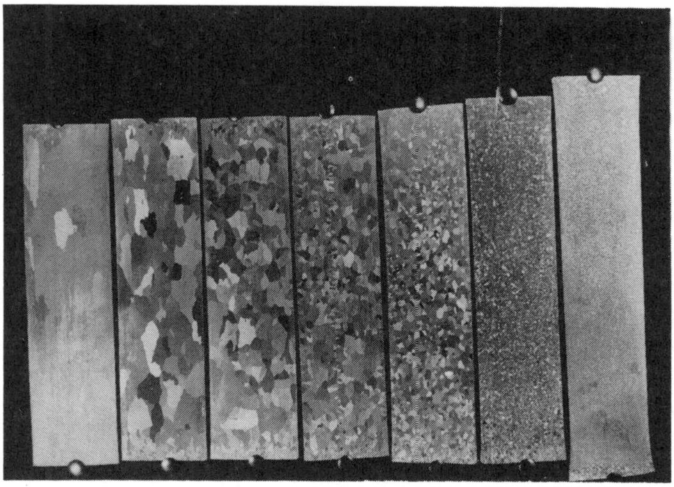

Figure 3.14 The effect on the grain size of annealing aluminium that has been subject to different amounts of cold work. Annealing involved, in all cases, 5 hours at 640°C. (From Rollason, E. C., *Metallurgy for Engineers,* Edward Arnold)

aluminium that had been subject to different amounts of cold work. The greater the amount of cold work the smaller the resulting grain size. The greater the amount of cold work the more centres are produced for crystal growth.

HOT WORKING

Cold working involves plastically deforming materials below the recrystallisation temperature. The result is a harder, less ductile, material with deformed grains. *Hot working* involves deforming a material at a temperature greater than the recrystallisation temperature. As soon as a grain becomes deformed it recrystallises, no hardening occurs and the working can be continued without any difficulty. No interruption of working is needed to anneal the material, as is the case with cold working.

The grain structure of hot-worked material depends on the temperature at which the working occurs, the type of working process involved and the cooling rate of the material after the hot working. If the temperature is just above the recrystallisation temperature a fine grain structure is produced, if it is well above this temperature then large grains will be produced.

A disadvantage of hot working is that oxidisation of the metal surfaces occurs; cold working does not have this problem. Another disadvantage is that the material will have comparatively low values of hardness and tensile strength, with high elongation. Hot and cold working processes are often both used in a particular shaping operation. The first operation involving large amounts of plastic deformation is carried out by hot working. After cleaning the material is then cold worked to increase the strength and hardness and give a good surface finish.

CAST AND WROUGHT PRODUCTS

Casting involves the shaping of a product by the pouring of liquid metal into a mould. The grain structure within the product is determined by the rate of cooling. Thus the metal in contact with the mould cools faster than that in the centre of the casting. This gives rise to small crystals, termed *chill crystals,* near the surfaces. These are small because the metal has cooled too rapidly for the crystals to grow to any size. The cooling rate nearer the centre is however much less and so some chill crystals can develop in an inward direction. This results in large elongated crystals perpendicular the mould walls, these being called *columnar crystals.* In the centre of the mould the

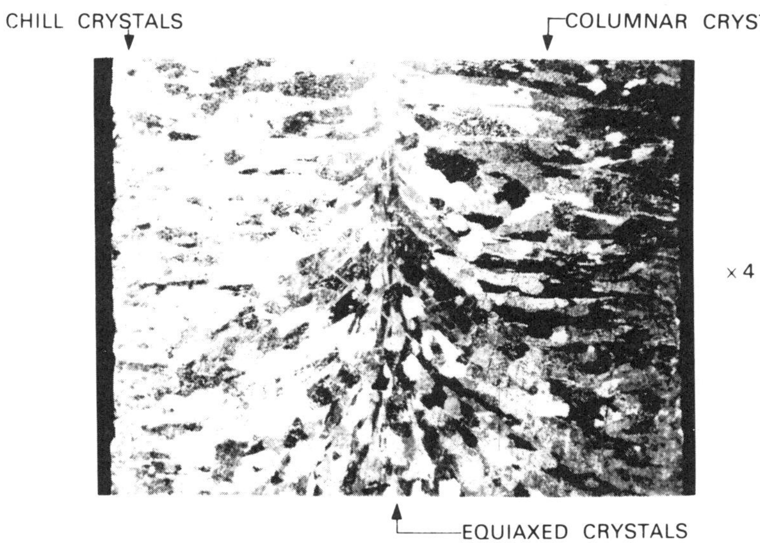

Figure 3.15 Grain structure in an aluminium casting. (From Monks, H. A and Rochester, D. C., *Technician Structure and Properties of Metals 2,* Cassell & Co.)

cooling rate is the lowest. While growth of the columnar crystals is taking place small crystals are growing in this central region. These grow in the liquid metal which is constantly on the move due to convection currents. The final result is a central region of medium-sized, almost spherical, crystals called *equiaxed crystals. Figure 3.15* shows all these types of crystals in a casting section.

In general a casting structure having entirely small equiaxed crystals is preferred. This type of structure can be promoted by a more rapid rate of cooling for the casting. Castings in which the mould is made of sand tend to have a slow cooling rate as sand has a low thermal conductivity. Thus sand castings tend to have large columnar grains and hence relatively low strength. Die casting involving metal moulds has a much faster rate of cooling and so gives castings having a bigger zone of equiaxed crystals. As these are smaller than columnar crystals the casting has better properties. The following table shows the types of differences that can occur with aluminium casting alloys.

Material	*Tensile strength*/N mm^{-2} *Sand cast*	*Tensile strength*/N mm^{-2} *Die cast*	*Percentage elongation Sand cast*	*Percentage elongation Die cast*
5% Si, 3% Cu	140	150	2	2
12% Si	160	185	5	7

Castings do not show directionality of properties, the properties being the same in all directions. They do however have the problems produced by working from a liquid metal of blowholes and other voids occurring during solidification.

Manipulative processes involve the shaping of a component by plastic deformation processes. The products given by such methods are said to be *wrought.* Hot working processes involve rolling, forging or extruding at temperatures in excess of the recrystallisation temperature. Cold working processes such as rolling, drawing, pressing, spinning and impact extrusion involve temperatures less than the recrystallisation temperature.

Rolling is a continuous process in which the material is passed between a pair of rotating rollers and emerges with a reduced thickness (as in *Figure 3.8*). Hot rolling is often at a temperature of about 0.6 T_m. At this temperature work hardening does not occur. The product is a relatively soft material with low tensile strength, the surfaces of the material are also oxidised. A ductile material can be cold rolled. Such a process work hardens the material and can lead to a useful gain in tensile strength and hardness. A good surface finish can be produced. Rolling does however give a directionality to the properties of the finished material due to the deformation of the grains in the direction of the rolling. The following table illustrates this for rolled brass strip (70% copper, 30% zinc).

Angle to rolling direction	*Tensile strength* /N mm^{-2}	*Percentage elongation*
0°	740	3
45°	770	3
90°	850	2

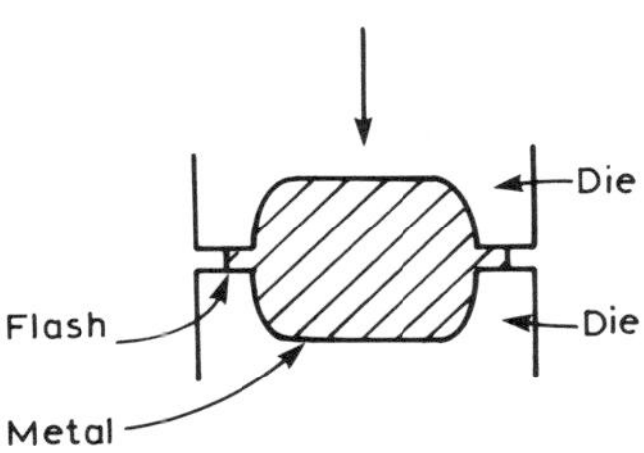

Figure 3.16 Closed-die forging

Forging involves squeezing a ductile material between a pair of dies. The term forging is generally only applied to the hot working process, though cold forging can be carried out with some of the very ductile non-ferrous materials. *Figure 3.16* shows the principle of one form of forging, closed-die forging. The flow of the material during the squeezing operation does give a directionality to the properties of the material.

Extrusion is rather similar to the squeezing of toothpaste out of the tube. The form of the ejected toothpaste is determined by the nozzle through which it is ejected. *Figure 3.17* shows the basic principles. Extrusion involves forcing a metal through a die. Hot extrusion involves temperatures of the order of 0.65 T_m to 0.9 T_m. The result of hot extrusion is a product with comparatively low tensile strength and soft. Cold extrusion gives a cold-worked product with higher strength and hardness. The process is not however possible where the cross-sectional area of the material is considerably reduced due to too much work hardening occurring. The process of extrusion does give a product with directionality of properties.

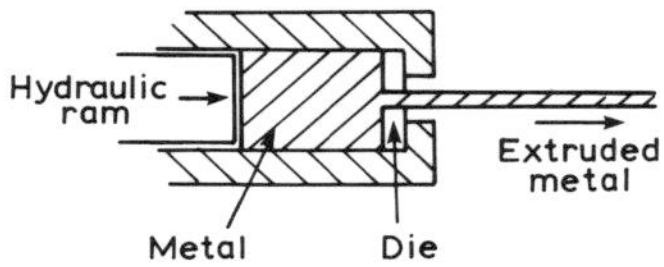

Figure 3.17 Extrusion

In general, wrought products have a directionality of properties due to the process used giving rise to grain deformation in some particular direction.

Example

Lead has a melting point of 327°C. Will the product be work hardened if lead is extruded at room temperature?

The melting point is about 600 K. This would mean that extrusion at about 300 K, about the order of room temperature is extrusion at about 0.5 T_m. This just qualifies as a hot-working process. Thus the product should show no work hardening. The recrystallisation temperature is about the same as room temperature.

THE EFFECT OF GRAIN SIZE ON PROPERTIES

Grain size is an important factor in determining the mechanical properties of a material. It is not so much the size of the grain that is important as the length of grain boundary. Fine-grain material contains a greater length of grain boundaries than a coarse-grained material. Grain boundaries restrict the amount of slip that can occur in that slip within one grain is not easily transmitted across a grain boundary to cause slip in a neighbouring grain. Thus the more grain boundaries there are the more slip is restricted. Coarse-grained materials thus have lower yield stresses than fine-grained materials. This also means that a coarse-grained material will have a lower tensile strength than a fine-grained material. Hardness is related to tensile strength and thus coarse-grained material will have a lower hardness than fine-grained material. Ductile materials are those which have a large plastic region in their stress–strain relationship (see page 6), thus we would expect the large-grain materials to be the more ductile as plastic deformation is easier.

The mechanical properties of a metal can be changed by changing the size or the shape of the grains. Cold-working distorts the shape of the grains and so increases tensile strength and hardness, while decreasing ductility. Annealing of a cold-worked material can increase ductility, while decreasing tensile strength and hardness, by reforming the grains without the distortion. It also enables the grain size to be controlled.

PROBLEMS

1. Distinguish between cold- and hot-working processes.

2. Describe the effect on the mechanical properties of a metal of cold working.

3. Explain what is meant by the recrystallisation temperature.

4. What is meant by 'work-hardening'?

5. What are the properties of a fully work hardened material in comparison with its properties before any working occurs?

6. How does cold working change the structure of a metal?

7. Describe how the mechanical properties of a cold-worked material change as its temperature is raised from room temperature to about 0.6 T_m, where T_m is the melting point temperature in degrees Kelvin.

8. What factors affect the recrystallisation temperature of a metal?

9. How is the recrystallisation temperature of a pure metal related to its melting point?

10. Zinc has a melting point of 419°C. Estimate the recrystallisation temperature for zinc.

11. Magnesium has a melting point of 651°C. What order of temperatures would be required to (a) stress relieve, (b) anneal a cold-worked piece of magnesium?

12. What is the effect on the grain size of a metal after annealing of the amount of cold work the material had originally been subject to?

13. How does the temperature at which hot working is carried out determine the grain size and so the mechanical properties?

14. Describe the grain structure of a typical casting.

15. Why are the properties of a material dependent on whether it is sand-cast or die-cast?

16. What is meant by directionality in wrought products?

17. Why are the mechanical properties of a rolled metal different in the direction of rolling from those at right-angles to this direction?

18. How does a cold-rolled product differ from a hot-rolled product?

19. Brasses have recrystallisation temperatures of the order of 400°C. Roughly what temperature should be used for hot extrusion of brass?

20. Describe how grain size and shape affect the mechanical properties of a metal.

21. A brass, 65% copper and 35% zinc, has a recrystallisation temperature of 300°C after having been cold-worked so that the cross-sectional area has been reduced by 40%.
   (a) How will further cold working change the structure and the properties of the brass?
   (b) To what temperature should the brass be heated to give stress relief?
   (c) To what temperature should the brass be heated to anneal it and give a relatively small grain size?
   (d) How would the grain size, and the mechanical properties, change if the annealing temperature used for (c) was exceeded by 100°C?

22. Use *Figure 3.9* for this question.
   (a) What is the maximum hardness possible with cold rolled copper?

(b) Copper plate, already cold worked 10% is further cold worked 20%. By approximately how much will the hardness change?
(c) Mild steel is to be rolled to give thin sheet. This involves a 70% reduction in sheet thickness. What treatment would be suitable to give this reduction and a final product which was no harder than 150 HV?

23. As a fine-grained product is generally desirable after some process, it is common to lower the temperature of a hot-working process to a working temperature just above the recrystallisation temperature for the final part of the process. Why should this lead to fine grains?

4 Steels

Objectives: At the end of this chapter you should be able to:
Explain the mechanical properties of plain carbon steels.
Explain the effects of adding alloying elements to plain carbon steels.
Define the critical points.
Explain and interpret time–temperature transformation diagrams and continuous cooling transformation curves.
Explain the effect of quenching steels in terms of the transformations occurring.
Explain the term 'hardenability' and interpret Jominy test results.
Explain the term 'limiting ruling section' and recognise the effect of the section size on the mechanical properties.
Describe the heat treatment processes of tempering, annealing, austempering, martempering and surface hardening and explain their effect on the properties of the steel concerned.
Describe the main coding systems used for steels and select steels using data given in specifications.

PLAIN CARBON STEELS

Plain carbon steels are essentially just alloys of iron and carbon. Other elements are however invariably present in such steels, e.g. manganese, silicon, sulphur and phosphorus. The term plain carbon steel is however restricted to those steels where the carbon does not exceed 1.5%, the manganese about 1.5%, the silicon about 0.5% and the other elements are no more than traces. The mechanical properties of such steels are determined primarily by the percentage of carbon in the steel.

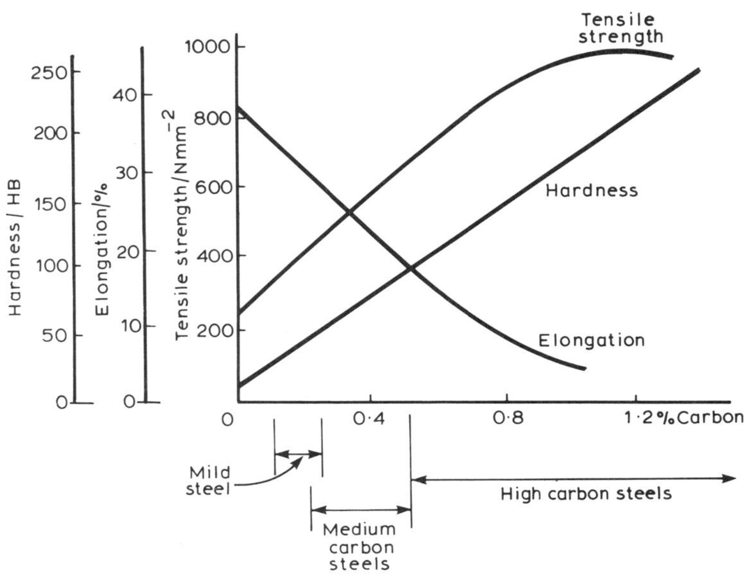

Figure 4.1 The mechanical properties of plain carbon steels

Plain carbon steels are often grouped into three categories. The term *mild steel* is used where the carbon content is between about 0.10 and 0.25%, *medium carbon steel* for between 0.20 and 0.50% carbon, and *high carbon steel* for between 0.5 and 1.5% carbon. *Figure 4.1* shows the effects of the carbon content on the mechanical properties of the steel. Increasing the carbon content increases the hardness and the tensile strength but decreases the elongation.

Mild steels have relatively low tensile strength and hardness but high ductility. Typical applications of mild steel are sections for use as joists in buildings, bodywork for cars and ships, screws, nails and wire. Medium carbon steels have higher tensile strength and hardness but lower ductility than mild steels. Typical applications are agricultural tools, fasteners, dynamo and motor shafts, crankshafts, connecting rods and gears. High carbon steels have the highest tensile strength and hardness but very low elongations. They are used where hardness is a more necessary requirement than ductility. Typical applications are tools, saws, hammers, chisels, punches, axes, dies, taps, drills and razors.

The mechanical properties of the plain carbon steels can be explained in terms of their microstructure. *Figure 4.2* shows the relevant part of the iron-carbon diagram (strictly the iron–iron carbide diagram) and the resulting microstructure of the steels when

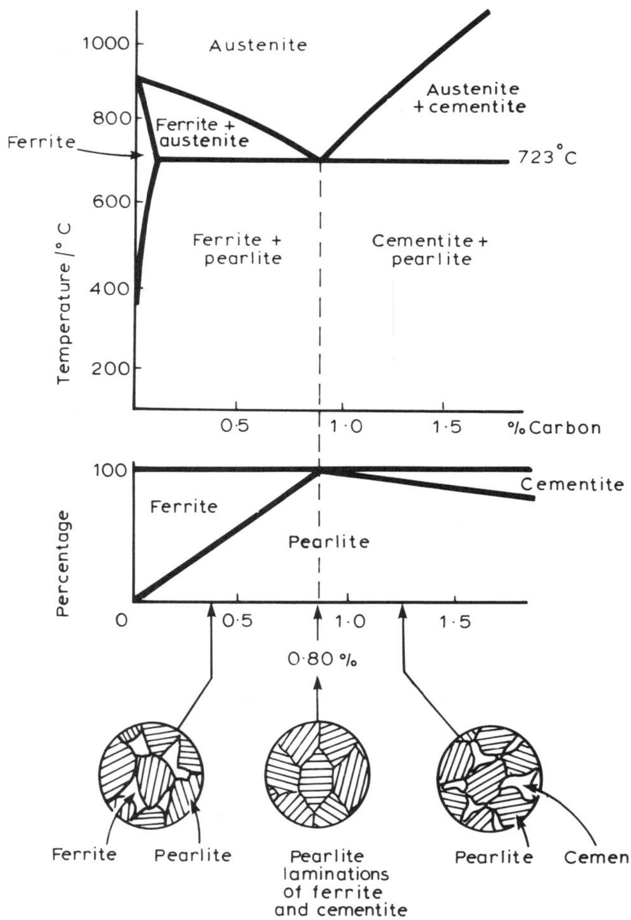

Figure 4.2 The microstructure of plain carbon steels

they are cooled slowly from the austenitic state. Up to a carbon content of 0.80%, the eutectoid composition, the microstructure consists of ferrite and pearlite. At 0.80% carbon the structure is entirely pearlite. For higher carbon percentages the structure consists of cementite and pearlite. *Ferrite* is soft and ductile, being a body-centred cubic solid solution of carbon in iron. *Pearlite* is stronger and harder than ferrite, being the eutectoid mixture of ferrite and cementite; it consists of alternate thin layers of ferrite and cementite. Thus as the percentage of ferrite decreases and the percentage of pearlite increases, for steels with carbon content increasing from 0 to 0.80%, so the tensile strength and hardness increase and the elongation decreases. For high percentages of carbon the microstructure involves pearlite and cementite. *Cementite* is iron carbide and is extremely hard and brittle. Thus increasing the amount of cementite increases the hardness of the steel and its brittleness.

Austenite is the face-centred cubic solid solution form which, in the case of plain carbon steels, does not occur below 723°C.

Plain carbon steels can be hardened by rapid cooling from the austenite state. This produces a very hard constituent called *martensite.* The rapid cooling is achieved by quenching in cold water. The resulting steel is hard and brittle, but its properties can be adjusted to a less brittle state by tempering. This involves reheating the steel to a temperature in the range of 200° to 300°C. The temperature used determines the final balance of hardness and brittleness, the lower the temperature the harder and the more brittle the material.

The limitations to the use of plain carbon steels are that high strength steel, up to about 600 N mm^{-2}, can be obtained only by increasing the carbon content to such a level that the material becomes brittle. Another problem is that high hardness is achieved by water quenching. The severity of this rapid rate of cooling is such as often to lead to distortion and cracking of the steel.

The preceding discussion of plain carbon steels is just an over-view of the subject. More details will be found in the preceding volume in this series *Materials Technology for Technicians 2.*

Example

You have a choice between two steels for particular application. One of the steels has a microstructure with 20% pearlite and 80% ferrite, the other has 80% pearlite and 20% ferrite. Which of the steels would be expected to have the higher tensile strength and hardness?

Ferrite is a soft ductile constituent of steel, while pearlite is stronger and harder. Thus the steel with the greater percentage of pearlite, i.e. the 80% pearlite – 20% ferrite steel, will have the higher tensile strength and hardness.

ALLOY STEELS

The term *alloy steel* is used to describe those steels to which one or more alloying elements, in addition to carbon, have been added deliberately in order to modify the properties of the steel. In general the steels have less than 1% carbon and the additions are, in the case of manganese, more than 1.5% or, in the case of silicon, more than 0.5%.

There are a number of ways in which alloying elements can have an effect on the properties of the steel:

(1) They may go into solid solution, giving an increase in strength. In *low-alloy steels,* i.e. those with less than about 3 to 4% of alloying elements other than carbon, the alloying elements enter into solid solution with the ferrite and so give a stronger steel which still has good ductility. Silicon is an example of such an alloying element.
(2) They may form stable, hard, carbides or nitrides which may, if in an appropriate form such as fine particles, increase the strength and hardness. Manganese, chromium and tungsten have this effect.
(3) They could cause the breakdown of cementite and lead to the presence of graphite in the structure. Silicon and nickel have this effect. The result is a decrease in strength and hardness. For this reason silicon and nickel are not used with high carbon steels.
(4) They may lower the temperature at which austenite is formed on heating the steel. Manganese, nickel, copper and cobalt have this effect. The lowering of this temperature means a reduction in the temperature to which the steel has to be heated for hardening by quenching. If a sufficiently high percentage of one of these elements is added to the steel the transformation temperature to austenite may be decreased to such an extent that the austenite is retained at room temperature. With manganese about 11 to 14% produces what is known as an *austenitic steel.* Such steels have relatively good hardness combined with ductility and so are tough.
(5) They may increase the temperature at which austenite is formed on heating the steel. Chromium, molybdenum, tungsten, vanadium, silicon and aluminium have this effect. This raises the temperature to which the steel has to be heated for hardening. If a sufficiently high percentage of one of these elements is added to the steel the transformation from ferrite to austenite may not take place before the steel reaches its melting point temperature. Such steels are known as *ferritic steels,* 12 to 25% of chromium with a 0.1% carbon steel gives such a type of steel. Such a steel cannot be hardened by quenching.
(6) They could change the critical cooling rate. The *critical cooling rate* is the minimum rate of cooling that has to be used if all the austenite is to be changed into martensite. With the critical cooling rate, or a higher rate of cooling, the steel has maximum hardness. With slower rates of cooling a less hard structure is produced. Most alloying elements reduce the critical cooling rate. The effect of this is to make air or oil quenching possible, rather than water quenching. It also increases the hardenability of steels (see page 54).
(7) They can influence grain growth. Some elements accelerate grain growth while others decrease grain growth. The faster grain growth leads to large grain structures and consequentially to a degree of brittleness. The slower grain growth leads to smaller grain size and so to an improvement in properties. Chromium accelerates grain growth and thus care is needed in heat treatment of chromium steels to avoid excessive grain growth. Nickel and vanadium decrease grain growth.
(8) They can affect the machinability of the steel. Sulphur and lead are elements that are used to improve the chip formation properties of steels.
(9) They can improve the corrosion resistance. Some elements promote the production of adherent oxide layers on the surfaces of the steel and so improve its corrosion resistance. Chromium is particularly useful in this respect. If it is present in a steel in excess of 12% the steel is known as *stainless steel* because of its corrosion resistance. Copper is also used to improve corrosion resistance.

The following list indicates the main effects of the commonly used alloying elements. As will have been apparent from the preceding discussion an alloying element generally has more than one way of affecting the properties of the alloy.

Element	*Typical amount*	*Main efforts on properties*
Aluminium	0.95 to 1.3%	Aids nitriding (see page 63). Restricts grain growth.
Chromium	0.5 to 2%	Increases hardenability.
	4 to 18%	Improves corrosion resistance.
Copper	0.1 to 0.5%	Improves corrosion resistance.
Lead	Less than 0.2%	Improves machinability.
Manganese	0.2 to 0.4%	Combines with the sulphur in the steel to reduce brittleness.
	1 to 2%	Increases hardenability.
Molybdenum	0.1 to 0.5%	Inhibits grain growth. Improves strength and toughness.
Nickel	0.3 to 65%	Improves strength and toughness. Increases hardenability.
Phosphorus	0.05%	Improves machinability.
Silicon	0.2 to 2%	Increases hardenability. Removes oxygen in steel making.
Sulphur	Less than 0.2%	Improves machinability.
Tungsten		Hardness at high temperatures.
Vanadium	0.1 to 0.3%	Restricts grain growth. Improves strength and toughness.

STRUCTURAL STEELS

Mild steel, with tensile strengths up to about 450 N mm^{-2}, or the high carbon steel, with a tensile strength of the order of 600 N mm^{-2}, has been widely used for structural work, axles to bridges. However the higher the percentage of carbon in the steel the more brittle the material became (see *Figure 4.1*). There thus is a problem of combining strength with toughness. This can be overcome by the use of alloy steels.

The properties required of a constructional steel are:

(1) High yield strength, this representing the maximum stress to which the material could be exposed in use.
(2) Tough. Not brittle.
(3) Good weldability.
(4) Good corrosion resistance.
(5) Low cost.

Composition	*Condition*	*Yield stress* /N mm^{-2}	*Tensile strength* /N mm^{-2}	*Percentage elongation*	*Izod value* /J
0.40% C, 0.90% Mn, 1.00% Ni.	Quenched & tempered at 600°C.	490	700	25	90
0.31% C, 0.60% Mn, 3.00% Ni, 1.00% Cr.	Oil quenched & tempered at 600°C.	820	930	23	104
0.40% C, 0.55% Mn, 1.50% Ni, 1.20% Cr, 0.30% Mo.	Oil quenched & tempered at 600°C.	990	1080	22	70

These properties can be promoted by the addition of nickel Unfortunately nickel promotes the breakdown of cementite to give graphite. Thus in a medium carbon steel chromium is also added as it promotes the formation of carbides. Such a nickel chromium steel can show brittleness after tempering, but this effect can however be reduced by the addition of a small percentage of molybdenum, about 0.3%. The above table shows the properties of typical steels.

TOOL STEELS

Tool steels require properties that permit their use for tools involved in cutting and shaping metals. This requires good wear resistance, hardness and generally toughness.

Plain carbon tool steels obtain their hardness from their high carbon content, between 0.5 and 1.5%. The steels need to be quenched in cold water to obtain maximum hardness. Unfortunately they are rather brittle when very hard, lacking toughness. Where medium hardness with reasonable toughness is required the carbon steel will have about 0.7% carbon. Where hardness is the primary consideration and toughness is not so important, a carbon steel with about 1.2% carbon may be used.

Alloy tool steels are made harder and more wear resistant by the addition to the steel of elements that promote the production of stable hard carbides. Manganese, chromium, molybdenum, tungsten and vanadium are examples of such elements. A manganese tool steel contains from about 0.7 to 1.0% carbon and 1.0 to 2.0% manganese. Such a steel is oil-quenched from 780 to 800°C and tempered between about 220 and 245°C. The manganese content may be replaced partially by chromium, such a change improving the toughness. An example of such a steel has 1.00% carbon, 0.45% manganese and 1.40% chromium. The steel is oil-quenched from about 810°C and tempered at 150°C. It is used for ball bearings, cams and instrument pivots.

Shock-resistant tool steels are designed to have toughness under impact conditions, e.g. hand and pneumatic chisels, punches, picks. For such properties, fine grain is necessary and can be achieved by the addition of vanadium.

Tool steels designed for use with hot working processes need to maintain their wear resistance and hardness at the temperature used. Chromium and tungsten when added to steels form carbides which are both stable and hard, resulting in steels which retain their properties to high temperatures. An example of such a steel has 0.3% carbon, 3% chromium, 10% tungsten, 0.3% vanadium, 0.3% molybdenum and a maximum of 0.3% manganese. The steel is oil-quenched from 1150°C and tempered at 570°C. It is used for dies in hot forging, drawing and extrusion.

Steels that are able to work effectively at high machining speeds are called high-speed steels. The high speed results in the tool becoming hot. High speed steels thus have to retain their properties at high temperatures, as high as 500°C. Such steels must not become tempered by the high temperatures produced during the machining. The combination of tungsten and chromium is found to give the required properties, the carbides formed by these elements being particularly stable at high temperatures. An example of such a steel has a 0.75% carbon, 4.1% chromium, 1.1% vanadium and 18%

tungsten. It is quenched in oil or an air blast from 1280°C and double tempered at 560°C (not a true tempering but a precipitation of tungsten/iron carbides and transformation of retained austenite to martensite). At 20°C the steel has a hardness of 875 HV and at 550°C this hardness has only dropped to 555 HV.

The following information is taken from the data in the information sheet of a manufacturer of tool steels. (Courtesy of Sanderson Kayser Ltd)

KE. 960 Shock Resisting Alloy Tool Steel

B.S. 4659 type BS1

Average analysis: Carbon 0.50%, Silicon 0.80%, Chromium 1.5%, Tungsten 2.25%, Vanadium 0.20%.

Features and uses

K.E. 960 is an alloy shock-resisting tool steel for both hot and cold work applications. The tungsten content of this steel confers fatigue resistance, the chromium content gives depth of hardness and resistance to abrasion.

K.E. 960 is suitable for cold work tools subject to heavy shock and uneven loading; for example, press tools used for punching heavy gauge material, shear blades, nut blanking tools, perforating and piercing punches.

The steel is also used with great success for chisels and punches required for heavy work on hard and tough materials.

K.E. 960 is resistant to heat checking. It is suitable for hot work applications where high fatigue strength in combination with medium hot hardness is desirable. Tools made from this steel may be water cooled in service with little risk of cracking. Typical hot work applications include mandrel bars for drawing steel tubes; hot heading, swaging, forming and gripper dies used in medium temperature work; punching, piercing and trimming dies and punches for medium temperature work; shear blades working at medium temperature including flying shear blades.

STAINLESS STEELS

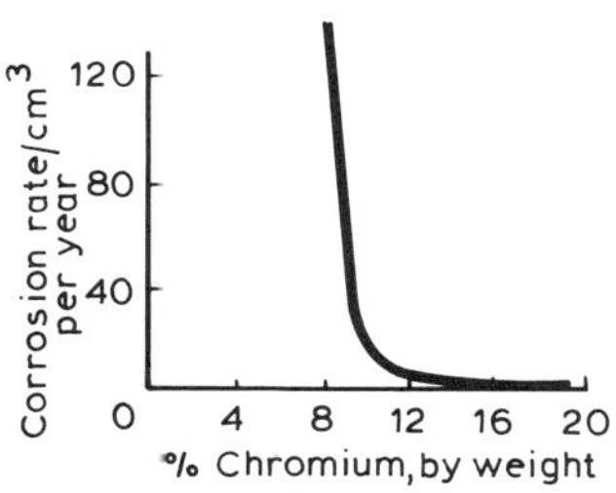

Figure 4.3 The effect of percentage of chromium on corrosion rate

Stainless steels contain a minimum of 10.5% chromium. The effect of the chromium is to confer exceptionally good corrosion resistance to the steel. *Figure 4.3* shows the effect of chromium content on the corrosion rate of a steel. The corrosion rate reaches a very low value when the chromium content is 12%. Stainless steels have excellent corrosion resistance, hence the term 'stainless'.

There are several types of stainless steels, ferritic, martensitic and austenitic. *Ferritic steels* contain between 12 and 25% chromium and less than 0.1% carbon. This high percentage of chromium with such a small percentage of carbon raises the temperature at which the transformation to austenite occurs to such a level that the steel remains in the ferritic state all the way up to its melting point. Because austenite cannot be formed with these steels, hardening by quenching to give martensite cannot occur. Martensite can only be produced by quenching from the austenite state. Thus such steels cannot be hardened by heat treatment. They can however be hardened by work hardening.

Martensitic steels contain between 12 and 18% chromium and between 0.1 and 1.5% carbon. The higher carbon content of these steels enables the austenite transformation to occur before the melting point of the steel is reached. Such steels can thus be quenched to give a martensitic structure.

Austenitic steels contain nickel as well as chromium. The effect of nickel is to lower the temperature at which austenite is produced, the reverse of the effect given by the chromium. A common steel of this type has 0.05% carbon, 18% chromium and 8.5% nickel. The transformation temperature is reduced by this combination to such an extent that the austenite is retained by the steel at room temperature. Such a steel cannot be hardened by quenching. They are however usually quenched, not to produce martensite, but to minimise the formation of chromium carbide as this causes a reduction in the corrosion resistance of the material.

The following table gives details of the mechanical properties of examples of the different types of stainless steels.

Ferritic steel: 0.06% carbon, 13% chromium.

Condition	*Yield stress* /N mm^{-2}	*Tensile strength* /N mm^{-2}	*Percentage elongation*	*Hardness* /HV
Annealed	280	416	20	170

Martensitic steel: 0.16% carbon, 12.5% chromium.

Condition	*Yield stress* / mm^{-2}	*Tensile strength* /N mm^{-2}	*Percentage elongation*	*Izod* /J	*Hardness* HV
Oil quenched from 1000°C	1190	1850	2.5	7	371
Oil quenched from 1000°C, tempered at 750°C.	370	570	33	134	172

Cutlery steel (a martensitic steel): 0.32% carbon, 13% chromium.

Condition	*Tensile strength* /N mm^{-2}	*Percentage elongation*	*Hardness* /HV
Oil quenched from 980°C, tempered at 180°C.	1450	8	600

Austenitic steel: 0.05% carbon, 18% chromium, 8.5% nickel, 0.8% manganese.

Condition	*Yield stress* /N mm^{-2}	*Tensile strength* /N mm^{-2}	*Percentage elongation*	*Hardness* /HV
Annealed	278	618	50	180
Cold rolled	803	896	30	

During welding, when temperatures of the order of 500° to 800°C are realised, structural changes may occur with the above steel. These changes are detrimental to the corrosion resistance of the material, the other properties being little affected. The effect is known as *weld decay* (which is in fact a form of intergrain corrosion). One way of overcoming this affect is to *'stabilise'* the steel by adding other elements such as niobium, titanium or columbium.

The following information is taken from the data in the information sheet of a manufacturer of stainless steel. (Courtesy of Arthur Lee & Sons Ltd.).

Stainless Types

(1) *Austenitic Chromium-Nickel Steels*

This group, typified by the well-known 18% chromium, 8% nickel quality, includes a great many variations, some containing molybdenum, while others are alloyed with columbium or titanium. Some are low in carbon, but each is designed for a specific purpose. Their structure is austenitic, a solid solution of chromium, nickel, iron and carbon, possessing the highest degree of corrosion resistance of any of the stainless family. Likewise they have high ductility and, while rapid cooling from high temperature puts them in their softest possible condition, they can be equally well cold work-hardened to high values of strength. Inherently they are tough and well adapted for fabrication by deep drawing, spinning, roll forming, etc., also readily welded by any known method, or soldered by proper techniques.

(2) *Martensitic Chromium Steels (Hardenable)*
This group includes the hardenable types such as the standard cutlery qualities and their variables with higher and lower carbons. These alloys, when cooled from a high temperature, develop greater tensile strength and hardness, and their structure is similar to that of a hardened carbon spring steel. By proper heat-treatment it is possible to develop a wide range of physical properties from 500 N mm^{-2} with 25% elongation, to over 1600 N mm^{-2} tensile with 2% to 4% elongation. Corrosion resistance of the alloys in this group is at its optimum in the heat-treated condition. Though not as corrosion resistant as the austenitic grades, it is adequate for the majority of engineering uses.

(3) *Ferritic Chromium Steels*
This last group is typified by the 16/18% chromium stainless steel; varieties are available containing aluminium and molybdenum which enhance respectively weldability and corrosion resistance. This steel does not harden when cooled rapidly from high temperatures. It is used generally in the fully-softened condition, in which state corrosion resistance achieves its maximum. The alloys in this group work-harden slowly which renders them suitable for fabrication by deep drawing, spinning, roll forming, etc., their structure being somewhat similar to that of a spheroidised carbon steel. Ferritic chromium steels have good oxidation resistance, although their anti-corrosive properties, while not quite equal to the austenitic group, are superior to those of martensitic steels.

In each of these three groups easy machining or freecutting qualities can be provided.

Selection of correct quality
Once it has been decided which group will satisfy the requirements of the application, details can then be studied and the proper modification of the general type chosen. For example, if the application requires forming by cold working, followed by welding, the part to be highly resistant to certain chemicals while exposed at times to temperatures between 450°C and 800°C, then firstly it would be decided that, from the welding point of view, the austenitic group is indicated. Because of the exposure to corrosive conditions after heating to certain temperatures, it is also apparent that a stabilised 18/8 quality would be required, leaving a choice of either 18/8 titanium stabilised, or 18/8 columbium stabilised. If the article has to be finally polished, the user would probably select the columbium stabilised. It is important that all conditions of fabrication and requirements of a new application be known in

advance, so that the proper type of stainless steel can be selected.

Some recommended applications:

(1) *Austenitic Steels*
Window and door frames. Roofing and guttering. Domestic hot water piping. Spoons, forks and knife handles. Kitchen utensils. Washing machines. Hospital equipment. Motor-car hub caps, rim embellishers and bumpers. Wheel spokes. Welding rods and electrodes. Wire ropes. Yacht fittings and masts and all marine applications. Nuts, bolts, screws, rivets, locking wire, split pins. Coiled and flat springs.

(2) *Martensitic Steels*
Flat springs. Coil springs. Scales and rulers. Knives. Spatulas. Kitchen tools and all appliances where high tensile and hardness is required with moderate corrosion resistance. Surgical and dental instruments. Record changer spindles.

(3) *Ferritic Steels*
Mouldings and trim for motor-car bodies, furniture, television sets, gas and electric stoves, refrigerators, etc. Coinage. Spoons and forks. Domestic iron soles. Vehicle silencers. Driving mirror frames. Nuts, bolts, screws, rivets, split pins. For heat-resistant applications such as oil-burner sleeves, or any part where working temperatures do not exceed 800°C.

MICROSTRUCTURAL TRANSFORMATIONS

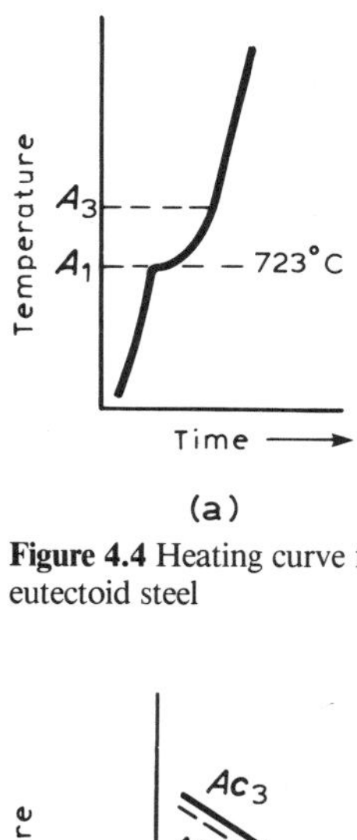

(a)

Figure 4.4 Heating curve for a hypo-eutectoid steel

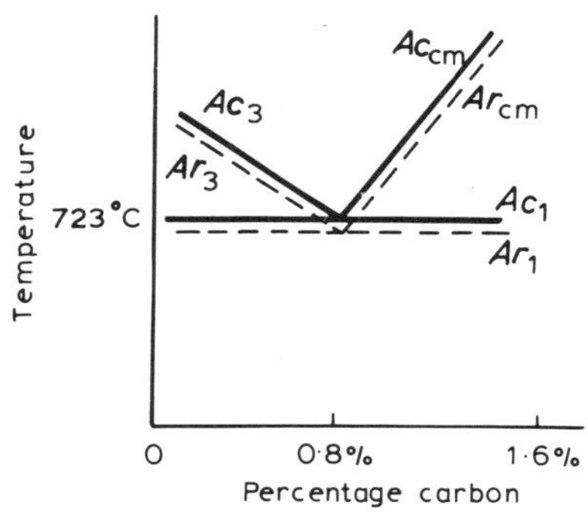

Figure 4.5 Critical points

The iron-carbon diagram shown in *Figure 4.2* shows the phases that will be present in iron-carbon alloys, of a range of compositions, at different temperatures if sufficient time is allowed for all the movements of atoms, i.e. diffusion processes, at any temperature to proceed to completion. If a steel is heated so that the temperature only rises slowly then the graph of temperature against time for the steel sample may be like that shown in *Figure 4.4*. At a particular temperature, in this case 723°C, the graph shows a discontinuity. Though heat is being supplied at a constant rate, the temperature does not rise at a constant rate at this particular temperature. Such a temperature is known as an *arrest point* or *critical point*. For the example given, a hypo-eutectoid steel, the temperature of 723°C marks the transformation of the steel from a ferrite-plus-pearlite structure to ferrite-plus-austenite. Critical points occur whenever the structure of the steel changes.

The lower critical point, i.e. 723°C, is denoted as A_1 and marks the transformation from a ferrite-plus-pearlite structure to one of ferrite-plus-austenite. The upper critical point is denoted as A_3 and marks the transformation from a ferrite-plus-austenite structure to one of just austenite.

For a hyper-eutectoid steel the lower critical point A_1 marks the transformation from a steel with a structure of pearlite plus cementite to one of austentite plus cementite. The upper critical point marks the transformation from the austenite-plus-cementite structure to austenite and is denoted by A_{cm}.

Cooling curves give critical points different slightly from those produced by heating, cooling giving lower values than heating. The cooling critical points are generally denoted by the inclusion of the letter ‘r’, e.g. Ar_1, and the heating points by the letter ‘c’, e.g. Ac_1. *Figure 4.5* shows the graph of the critical point temperatures plotted against the percentage of carbon in the steel. The graph is, in fact, essentially just the iron-carbon diagram of *Figure 4.2*.

The time scales for the iron-carbon diagram in *Figure 4.2* or the critical points in *Figure 4.5* is relatively long; long enough for all diffusion processes to be completed, i.e. reach equilibrium. Faster temperature changes can however lead to the diffusion processes not being completed, i.e. not reaching equilibrium. The faster changes can thus lead to the microstructure being in a non-equilibrium state, i.e. not that forecast by the iron–carbon diagram.

The effect of time and temperature on the microstructure of a steel can be expressed in terms of a *time-temperature transformation (TTT) diagram*. Such a diagram is sometimes known as an *isothermal transformation (IT) diagram* or *Bain S curves*. The diagram shows the rate at which austenite is transformed, at a given temperature, and the products produced by the transformation and any further transformations. Such a diagram is specific to just one particular steel. Thus steels with different percentages of carbon give different TTT diagrams, as do steels in which the other alloying elements are different.

One way of obtaining a TTT diagram is, for a particular steel, to take a small sample of the steel and heat it to a temperature at which it is completely transformed into austenite. The sample is then quickly transferred to a bath of liquid at some predetermined temperature. The time for which the sample is at this temperature is measured. This is the time during which the austenite transformation is occurring and the temperature of this change is that of the bath of liquid. This austenite transformation is taking place at a constant temperature, hence the term isothermal transformation (isothermal means constant temperature).

After the required time the sample is quickly quenched in cold water. This halts the transformation that was occurring and has the effect of converting the remaining austenite into martensite. The amount of martensite produced can be determined by microscopic

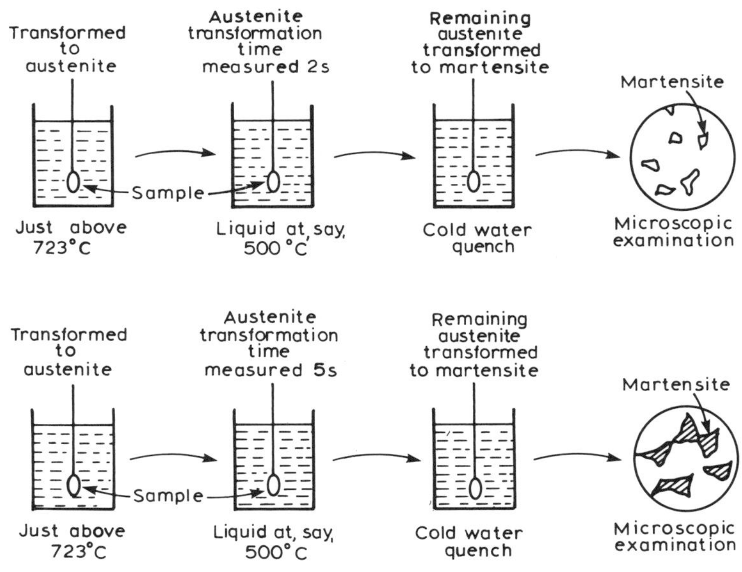

Figure 4.6 Obtaining two points on the TTT diagram at one particular temperature, 500°C

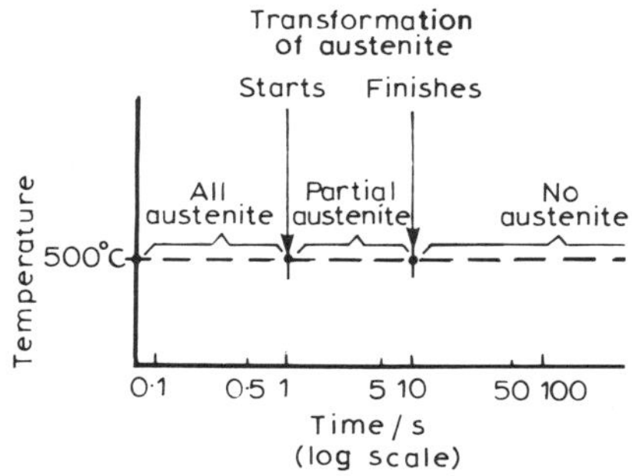

Figure 4.7 The transformation at one particular temperature

examination of the sample. Hence the amount of the sample that had not been transformed from austenite, at the temperature of the bath, can be determined. This type of investigation is carried out for a number of different times at the particular bath temperature (*Figure 4.6*). The way in which the microstructure of the steel changes with time can be represented by the line drawn in *Figure 4.7.* By repeating the above sequence of operations for different temperatures a composite diagram can be built up, the TTT diagram (*Figure 4.8*).

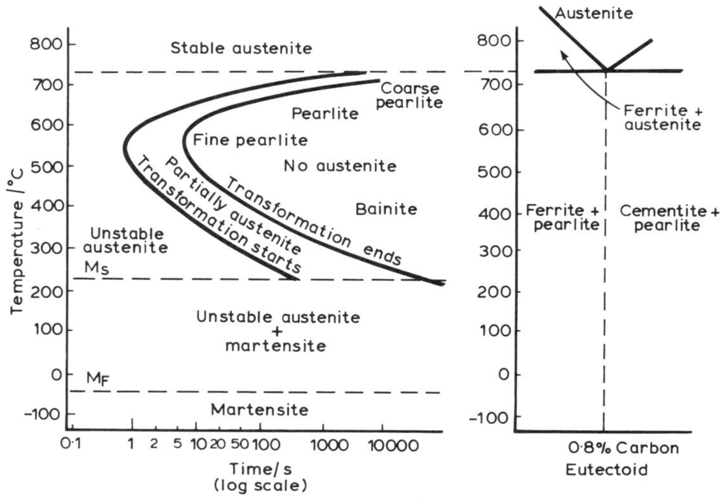

Figure 4.8 The TTT diagram for a 0.8 per cent plain carbon steel and the thermal equilibrium diagram

Above 723°C, for the steel chosen for *Figure 4.8*, austenite is the only stable phase. For temperatures below this the austenite is unstable and given sufficient time suffers a transformation. From 723°C down to about 566°C, the lower the temperture the shorter the time taken for the austenite to begin to transform. On completion of the transformation the microstructure is pearlite, the higher temperature giving a coarser pearlite than the lower temperature. From 566°C down to about 215°C the time taken for the austenite to begin to transform increases. The product of the transformation is known as *bainite.* Bainite, like pearlite, is a mixture of ferrite and cementite, however while pearlite consists of alternate layers of the two constituents, bainite consists of a feathery-like structure. From 215°C down to -20°C the unstable austenite

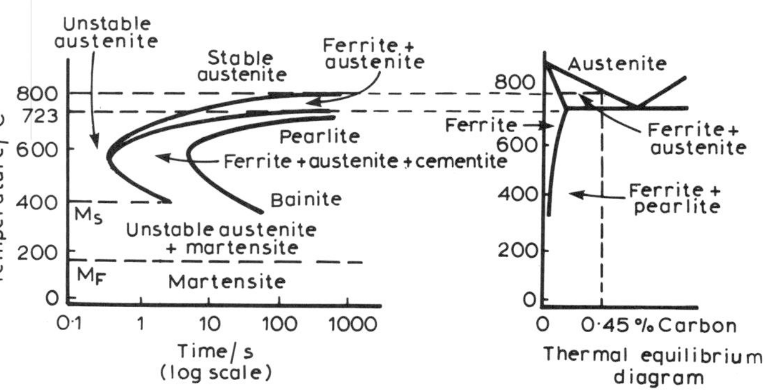

Figure 4.9 The TTT diagram for a 0.45 per cent plain carbon steel and the thermal equilibrium diagram

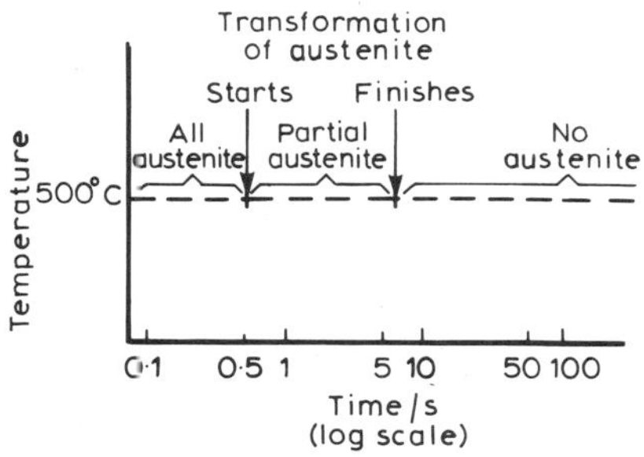

Figure 4.10 The transformations at one particular temperature for a 0.45 per cent plain carbon steel

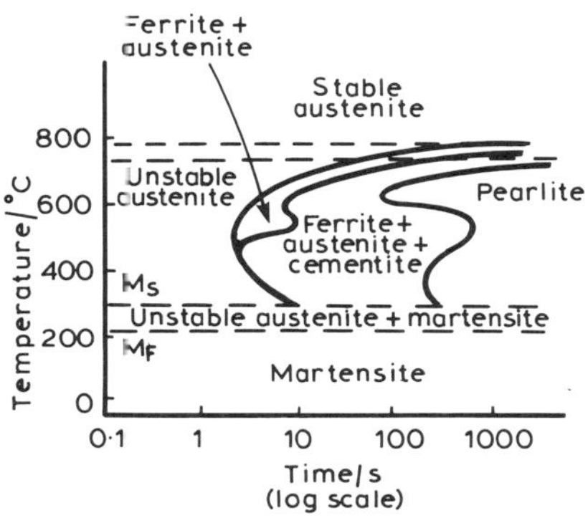

Figure 4.11 The TTT diagram for a 0.4 per cent carbon, 0.8 per cent chromium steel

transforms to martensite. At this lower temperature the structure is just martensite. For comparison with the conditions which would occur with very slow cooling, the relevant part of the thermal equilibrium diagram is also included in *Figure 4.8*.

In the region 215°C to −20°C the amount of martensite formed is practically independent of time, being determined mainly by the temperature at which the steel is. The temperature at which martensite begins to form is denoted by M_s and the temperature at which the transformation is complete by M_f.

The TTT diagram in *Figure 4.8* is for the eutectoid composition, i.e. 0.8% carbon. Different TTT diagrams are produced for other carbon contents. *Figure 4.9* shows the diagram for a 0.45% carbon steel, also the relevant part of the thermal equilibrium diagram. The TTT diagram for the 0.45% carbon steel differs from that of the 0.8% carbon steel in that there is an additional region due to ferrite transformations. Also, with the lower percentage carbon the TTT diagram has effectively been shifted to the left, i.e. the transformations begin more quickly and are completed more quickly at any given temperature. For example, compare the transformations at 500°C for the 0.45% carbon steel (*Figure 4.10*) with those given in *Figure 4.7* for the 0.8% carbon steel.

The lower the percentage of carbon in a plain carbon steel the more to the left the TTT diagram is shifted. This means a shorter time is needed at any particular temperature for the transformation to pearlite, or bainite.

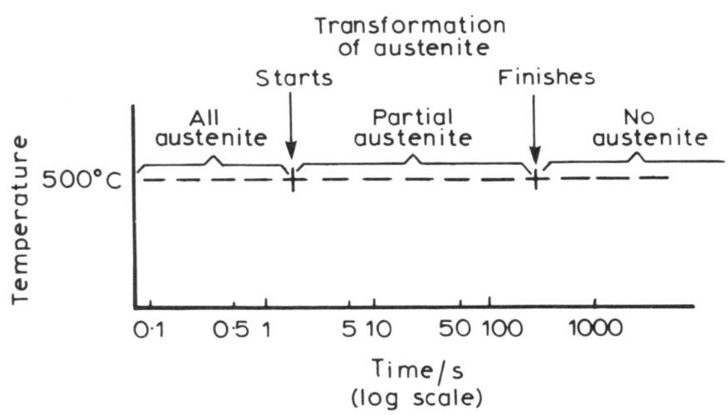

Figure 4.12 The transformation at one particular temperature for a 0.4 per cent carbon, 0.8 per cent chromium steel

The TTT diagram is also changed by other elements added to a carbon steel. *Figure 4.11* shows the TTT diagram for an alloy steel having 0.4% carbon and 0.8% chromium. The effect of the chromium is to shift the TTT diagram to the right, i.e. increase the time needed at any particular temperature for the transformation to pearlite or bainite (compare *Figure 4.12* with *Figure 4.10*). Nickel, molybdenum and manganese all have similar effects to that shown by the chromium additive to the steel.

QUENCHING STEELS

A hard, strong, steel can be produced by heating the steel to a temperature greater than the A_3 critical temperature so that the microstructure is entirely changed to austenite, then cooling quickly so that the austenite is transformed into martensite. The treatment thus involves continuous cooling. The cooling rate depends on the medium used for the quenching, water giving a faster cooling rate than oil.

The TTT diagrams are derived from data obtained for transformations at constant temperature. However, the diagrams are only slightly modified if continuous cooling is considered, the effect being to displace the TTT diagram to the right and downward. Also the continuous curves frequently show gaps where no transformation occurs on cooling. Such a diagram is referred to as a *continuous cooling transformation curve.*

Figure 4.13 shows the continuous cooling transformation curve for a 0.8% carbon steel, i.e. the eutectoid composition. Superimposed on the diagram are graphs of temperature against time that could occur

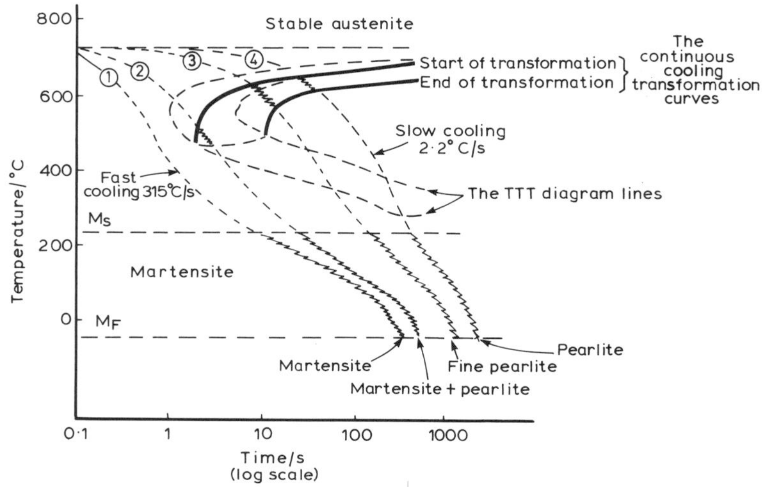

Figure 4.13 Cooling rate graphs superimposed on the continuous cooling transformation curves for a 0.8 per cent plain carbon steel (indicates where transformations are taking place)

for the steel when quenched in different media, i.e. different cooling rate graphs. With fast cooling (curve 1) from the stable austenite temperature the only transformation that occurs is from the austenite to martensite. With a slightly slower cooling rate (curve 2) a partial transformation to pearlite occurs and the final microstructure is thus martensite and pearlite. With an even slower cooling rate (curve 3) the austenite changes entirely to pearlite and the final structure is fine pearlite, no martensite being produced. With the yet slower cooling rate (curve 4) the austenite changes entirely to pearlite and again no martensite is produced.

For the microstructure to be entirely martensite, and hence show the maximum hardness, the cooling rate must be such that the only transformation occurring is from austentite to martensite. For the structure to be entirely pearlite the cooling rate must be such that the cooling rate line passes through the start and end of transformation lines. The structure will be partly pearlite and partly martensite if the cooling rate line passes into the transformation region but does not pass completely through it from the start to end of transformation lines.

If a steel is being quenched to give an entirely martensitic microstructure then the cooling rate line must not pass through the transformation region. The minimum cooling rate that will do this is

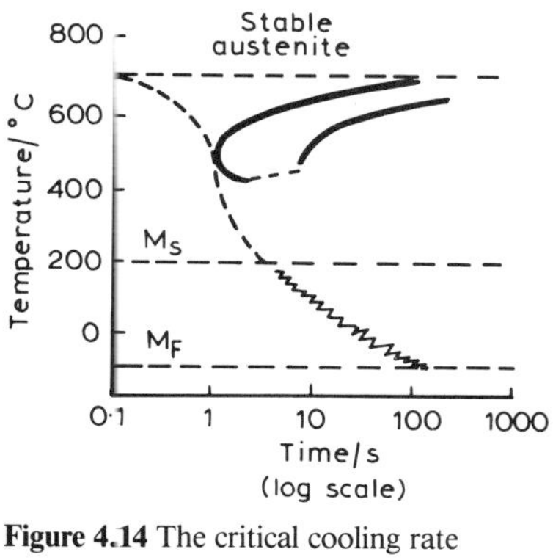

Figure 4.14 The critical cooling rate

shown in *Figure 4.14.* This minimum cooling rate is called the *critical cooling rate.* Cooling rates faster than this will give a completely martensite structure but rates slower than this will not.

The critical cooling rate thus depends on the form of the continuous cooling transformation curve (and hence the TTT diagram). *Figure 4.9* shows the TTT diagram for a 0.45% plain carbon steel. The result of having the lower percentage of carbon, than that giving *Figure 4.13*, has been to shift the diagram to the left. This means that the critical cooling rate for the 0.45% carbon steel is higher than that for the 0.8% carbon steel. The more the TTT diagram is shifted to the left the faster the cooling rate needed to obtain a completely martensitic structure. In fact, by about 0.3% carbon the TTT diagram has been shifted so far to the left that the cooling rate required for such a change is too fast to be possible and so such steels cannot be hardened by quenching.

A problem with the quenching of plain carbon steels to give a martensitic structure is that the high cooling rates needed inevitably mean water quenching. But such rapid quenching can lead to cracking and distortion. The effect of adding elements, other than carbon, to steel is to shift the TTT curve, and hence the continuous cooling transformation curve, to the right. This means that the critical cooling rate is less and thus oil or salt quenching, or even air cooling, may be used and still result in a martensitic structure.

The rate of cooling that occurs for a piece of steel when quenched in, say, water depends on the size of the piece of steel. The bigger the piece of steel the slower its rate of cooling. The quenching medium has thus to be chosen with a particular size piece of steel in mind. *Figure 4.15* shows the continuous cooling transformations that occur for specific quenching media as a function of the diameters of the bars being quenched. Thus for a 10 mm diameter bar, air cooling will result in a microstructure involving ferrite, pearlite and a small

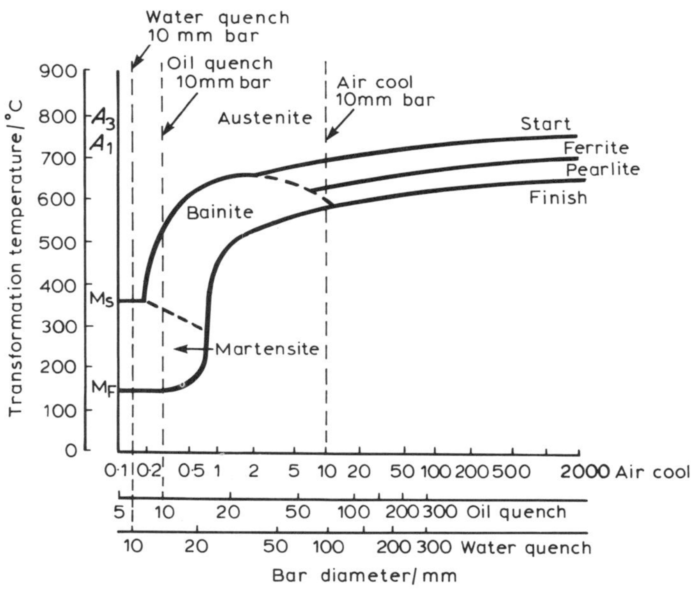

Figure 4.15 Continuous cooling transformations as a function of bar diameter and quenching media, for a 0.38 per cent carbon – 0.70 per cent manganese plain carbon steel. (Based on a diagram in *Atlas of Continuous Cooling Transformation Diagrams for Engineering Steels,* British Steel Corporation, 1977)

amount of bainite. With oil quenching the result is bainite plus martensite. With water quenching the microstructure is only martensite. Thus for this particular diameter bar water quenching must be used if a completely martensitic structure is required. The microstructures given by the diagram are those at the centre of the quenched bar and those at the surface may differ due to the surface cooling at a different rate.

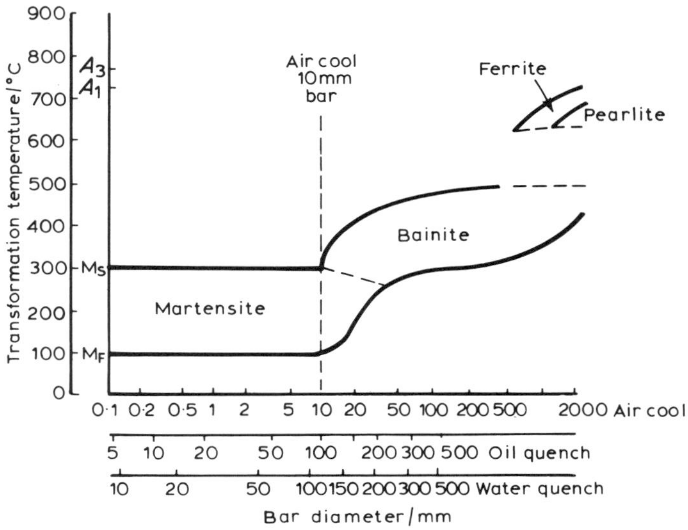

Figure 4.16 Continuous cooling transformations as a function of bar diameter and quenching media, for a 0.40 per cent carbon – 1.50 per cent nickel – 1.20 per cent chromium – 0.30 per cent molybdenum alloy steel. (Based on a diagram in *Atlas of Continuous Cooling Transformation Diagrams for Engineering Steels,* British Steel Corporation, 1977)

Figure 4.15 was for a plain carbon steel, *Figure 4.16* is however for a steel with a similar percentage of carbon but with chromium and molybdenum alloying elements. As will be apparent from a comparison of the two diagrams, a much less severe quenching can be used and still a martensitic structure obtained. For the 10 mm diameter bar air cooling will suffice.

Example

Which of the steels giving the continuous cooling transformation curves in *Figure 4.17* can be quenched in water and give a completely

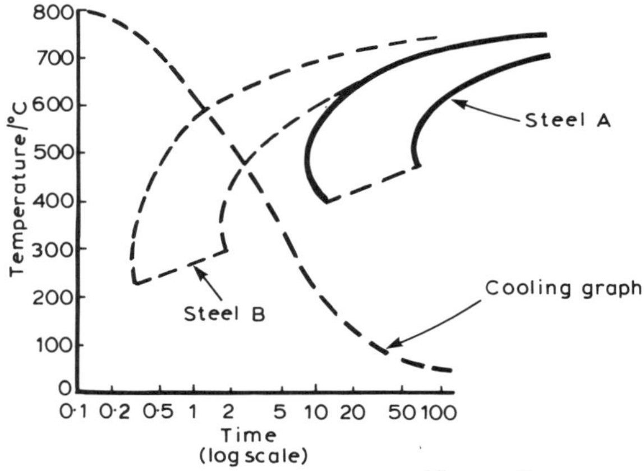

Figure 4.17 Continous cooling transformation curves for two steels

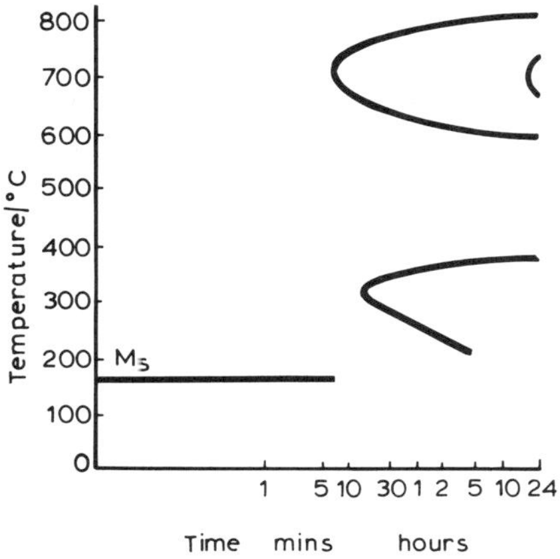

Figure 4.18 The TTT diagram for K.E.A. 162 steel (1.0 per cent carbon, 0.65 per cent manganese, 5.0 per cent chromium, 1.0 per cent molybdenum, 0.20 per cent vanadium). (Courtesy of Sanderson Kayser Ltd)

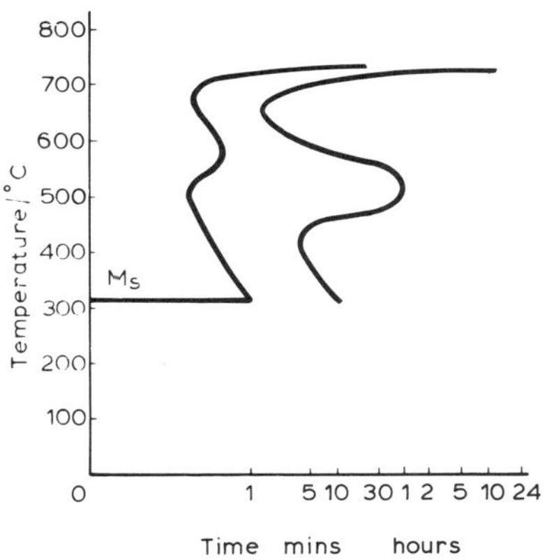

Figure 4.19 The TTT diagram for K.E. 896 steel (0.50 per cent carbon, 0.65 per cent manganese, 1.1 per cent chromium, 0.20 per cent vanadium). (Courtesy of Sanderson Kayser Ltd)

martensite structure if the cooling rate possible with the samples of the steels is given by that indicated in the figure?

Steel 'A' is the one that will completely transform from austenite to martensite without undergoing any other transformation. Steel 'B' will give a pearlite structure.

Example
Figure 4.18 is the TTT diagram for a cold work tool steel (K.E. 162) which is hardened by either air cooling or an oil quench. *Figure 4.19* is the TTT diagram for a medium carbon chrome vanadium steel (K.E. 896) which is hardened by quenching in oil except in the case of sections thicker than about 60 mm which require quenching in water to achieve maximum hardness. How can an examination of the TTT diagrams disclose the fact that the K.E. 162 steel requires a slower rate of cooling than the K.E. 896 steel? Why do the thicker sections of K.E. 896 require a faster rate of cooling than thinner sections?

The TTT diagram for the K.E. 162 steel is further to the right than the K.E. 896 steel and hence has a slower critical rate of cooling. Thus a fully hardened structure may be realised with air cooling or an oil quench while the K.E. 896 steel requires at least an oil quench if not a water quench. The thicker the section the longer it takes for the central regions of the section to come to the temperature of the surface. This means that a faster rate of cooling is needed for the surface in order that the central regions will not be below the minimum rate of cooling that is needed for a fully hardened structure, i.e. the critical cooling rate.

Example
Use *Figure 4.15* and *Figure 4.16* for this question. What microstructures are produced for the plain carbon and alloy steels specified as giving the continuous cooling transformation diagrams under these conditions?

(a) 50 mm diameter bar of the plain carbon steel is air cooled.
(b) 20 mm diameter bar of the plain carbon steel is water quenched.
(c) 50 mm diameter bar of the alloy steel is air cooled.
(d) 20 mm diameter bar of the alloy steel is water quenched.

(a) Ferrite plus pearlite, (b) Bainite plus a small amount of martensite, (c) Bainite, (d) Martensite.

HARDNESS

The maximum hardness that can be produced with any carbon steel is that occurring when the steel has a completely martensitic structure. *Figure 4.20* shows how the hardness of such a structure depends on the percentage of carbon in the steel. It also shows the hardness that would be produced if the steel had been given a slow quench, air cooled, and an entirely pearlitic structure had been produced.

When a carbon steel is quenched to room temperature, though the cooling rate may be fast enough to convert all the austenite to martensite, the entire transformation may be incomplete. The temperature at which austenite begins to transform to martensite M_s depends on the carbon content of the steel.

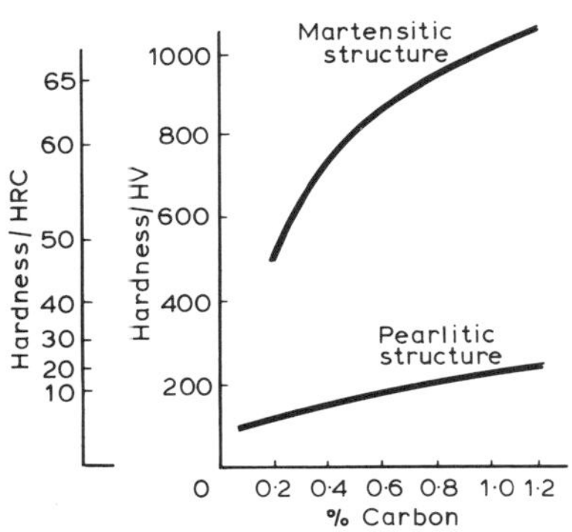

Figure 4.20 The hardness of carbon steels with martensitic or pearlitic structures as a function of carbon content

Carbon content	M_s/°C
0.2	490
0.4	420
0.8	250
1.2	150

The temperature at which the transformation from austenite to martensite is complete M_F is about 215°C below the M_s temperatures. Thus for the 0.2% carbon steel the martensite will be complete by 275°C, the 0.8% carbon steel by 35°C. Both these are above room temperature. However the M_F temperature for the 1.2% carbon steel is −65°C and thus such a steel will consist at room temperature of martensite plus austenite. This has the effect of reducing the overall hardness of that steel to below that it would have attained if completely martensitic.

HARDENABILITY

When a block of steel is quenched the surface can show a different rate of cooling to that of the inner core of the block. Thus the formation of martensite may differ at the surface from the inner core. This means a difference in hardness. *Figure 4.21* shows how the hardness varies with depths for a number of different diameter bars when quenched in water and in oil.

With the water quenching, the hardness in the inner core is quite significantly different from the surface hardness. The larger diameter bars also show lower surface and core hardness as their increased mass has resulted in a lower overall rate of cooling. With the oil quenching the cooling rates are lower than those of the water-cooled bars and thus the hardness values are lower. The hardness value for a fully martensitic 0.48% plain carbon steel is about 60 HRC. Thus, as the values in *Figure 4.21* indicate, the quenched bars are not entirely martensitic.

The term *hardenability* is used as a measure of the depth of

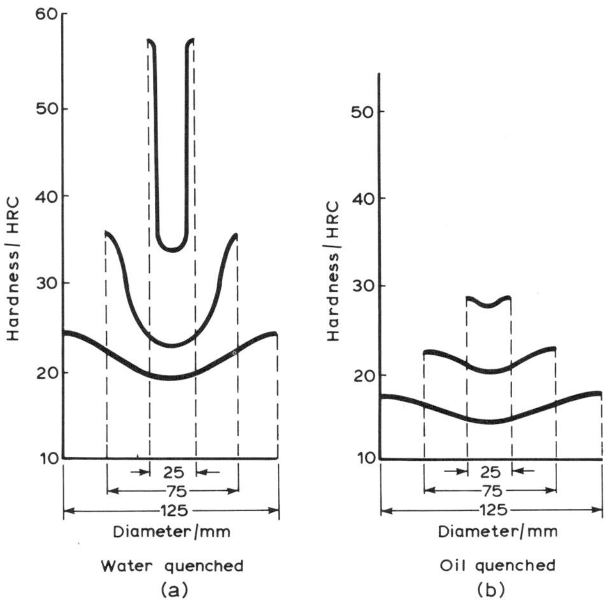

Figure 4.21 Variation of hardness with depth in quenched bars of different diameters, for a 0.48 per cent plain carbon steel

hardening introduced into a steel section by quenching (not to be confused with hardness). Hardenability is measured by the response of a steel to a standard test. The *Jominy test* involves heating a standard test piece of the steel to its austenitic state, fixing it in a vertical position and then quenching the lower end by means of a jet of water (*Figure 4.22*). This method of quenching results in different

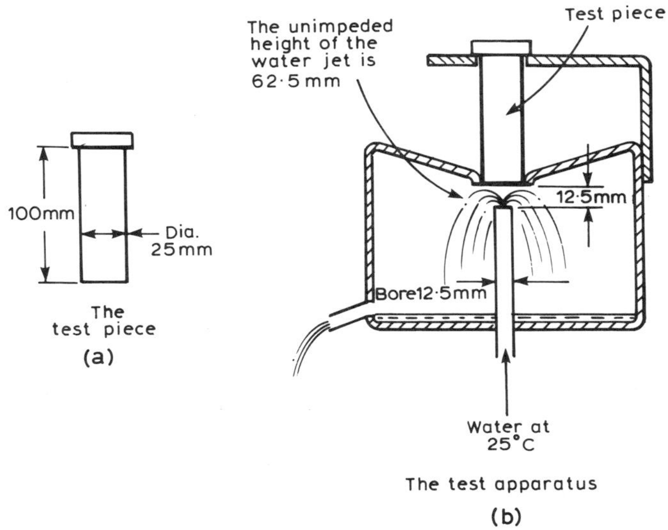

Figure 4.22 The Jominy test (BS 4437)

rates of cooling along the length of the test piece. When the test piece is cool, after the quenching, a flat portion is ground along one side of the test piece, about 0.4 mm deep, and hardness measurements made along the length of the test piece. *Figure 4.23* shows the types of result that are produced. The hardness is greatest at the end on which the cold jet of water played and least at the other end of the bar.

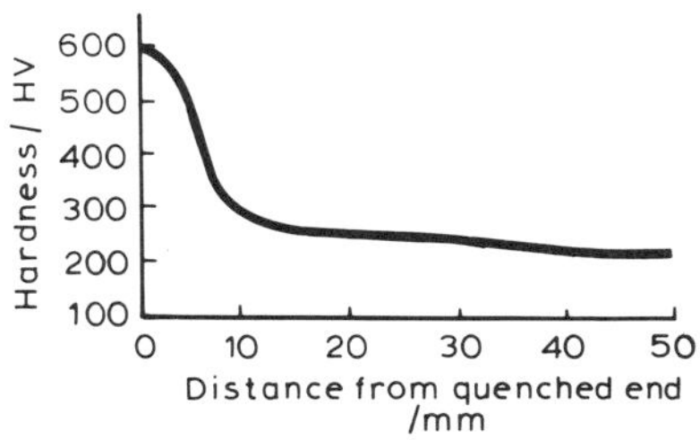

Figure 4.23 Results of the Jominy test for a 0.40 per cent plain carbon steel

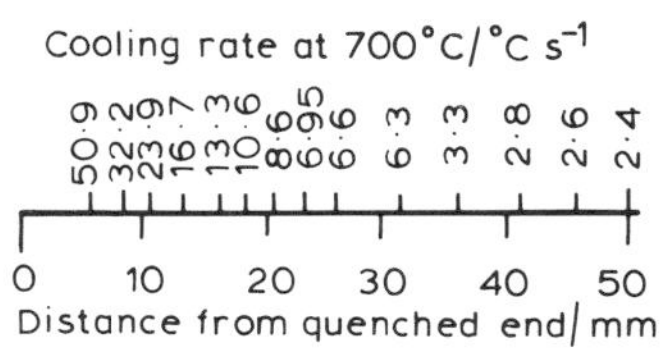

Figure 4.24 Cooling rates at different distances from the quenched end of the Jominy test piece

The significant point about the Jominy test results is not that they give the hardness at different distances along the test piece but that they give the hardness at different cooling rates. Each distance along the test piece corresponds to a different rate of cooling (*Figure 4.24*). The important point however is that this applies to both points on the surface and points inside any sample of the steel, provided we know the cooling rates at those points. This applies regardless of the quenching medium used. *Figure 4.25,* for example, shows how the cooling rate at the centre of circular cross-section bars at 700°C is related to their diameter. As the cooling rates are related to distances along the Jominy test piece these distances are also given in the figure.

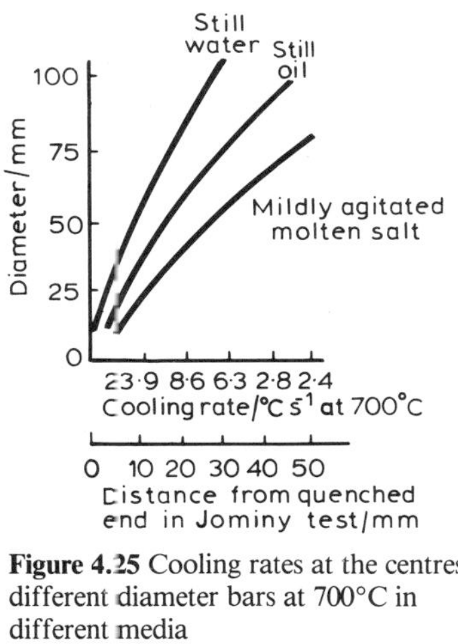

Figure 4.25 Cooling rates at the centres of different diameter bars at 700°C in different media

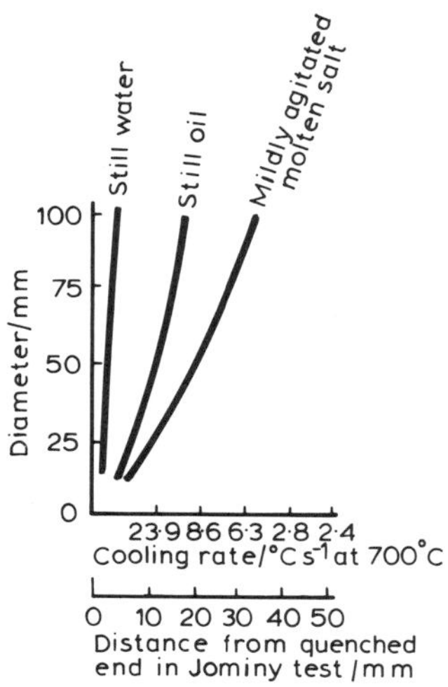

Figure 4.26 Cooling rates at the surfaces of different diameter bars at 700°C in different media

Thus for the Jominy test result given in *Figure 4.23,* the hardness at 20 mm from the quenched end of the test piece is about 230 HV. This means that, using *Figure 4.25,* a circular cross-section bar of diameter about 75 mm would have this hardness at its centre when quenched in still water from 700°C. If still oil had been the quenching medium the hardness at the centre of a 60 mm diameter bar would have been 230 HV.

Figure 4.26 shows, in a similar manner to *Figure 4.25,* how the cooling rate is related to the diameter of circular cross-section bars for points on the surface. The Jominy test result of 230 HV at 20 mm from the quenched end of the test piece means, using *Figure 4.26* , a circular cross-section bar of diameter 50 mm would have this hardness at its surface when quenched from 700°C in mildly agitated molten salt.

Figure 4.27 shows Jominy test results for two different steels. In order to enable the significance of the results to be seen in terms of

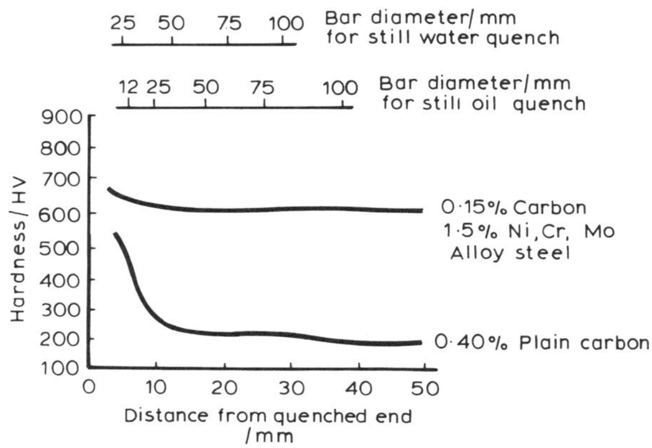

Figure 4.27 Results of Jominy tests

the hardness at the centre of different diameter bars, the cooling rates have been transposed into diameter values by means of *Figure 4.25.* The alloy steel can be said to be have better hardenability than the plain carbon steel. The plain carbon steel cannot be used in a diameter greater than about 25 mm with a still water quench if the bar is to be fully hardened. However the alloy steel is fully hardened for bars with diameters in excess of 100 mm.

HARDENABILITY AND CONTINUOUS COOLING CURVES

When a bar of material is cooled the centre of the bar undergoes a slower rate of cooling than the surface. The greater the diameter of the bar the greater the difference between these two rates of cooling.

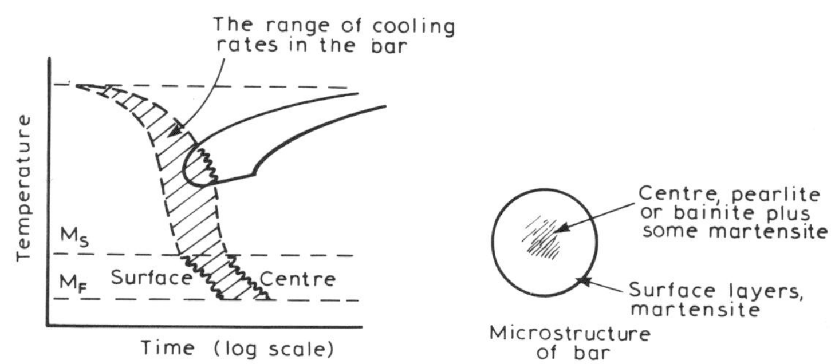

Figure 4.28 Quenching a bar

Figure 4.28 shows how the spread of cooling rates between the centre of a bar and its surface can be represented on a continuous cooling transformation curve in order to determine the structure across the section. For the example given in *Figure 4.28* the central region of the bar has been partially transformed to pearlite or bainite and thus is not as hard as the surface layers which have been completely transformed to martensite. The greater the cross-section size of the bar the greater the spread of cooling rates and thus the more difficult it is to completely harden the bar.

RULING SECTION

If you look up the mechanical properties of a steel in the data supplied by the manufacturer or other standard tables you will find that different values of the mechanical properties are quoted for different limiting ruling sections. The *limiting ruling section* is the maximum diameter of round bar at the centre of which the specified properties may be obtained. Here is an example of such information:

Steel	*Condition*	*Limiting ruling section*/mm	*Tensile strength* /N mm^{-2}	*Minimum elongation* %
070 M 55	Hardened	19	850 to 1000	12
	and	63	770 to 930	14
	tempered	100	700 to 850	14

The reason for the difference of mechanical properties for different size bars of the same steel is that during the heat treatment different rates of cooling occur at the centres of such bars due to their differences in sizes (see *Figure 4.21*). This results in differences in microstructure and hence differences in mechanical properties.

TEMPERING

Martensite is a hard and brittle substance. Thus a steel which has been quenched so that its entire microstructure is martensitic will be not only hard but probably very brittle, but it will not be tough. The toughness of such a steel can be improved by tempering. The gain in ductility is however balanced by a reduction in strength and hardness.

Tempering is the name given to the process in which a steel, hardened as a result of quenching, is reheated to a temperature below the A_1 temperature in order to modify the structure of the steel. Martensite is a highly stressed supersaturated solid solution of carbon in a distorted ferrite lattice. Heating such a structure enables the carbon atoms trapped in the ferrite lattice to diffuse out of that lattice and form fine cementite. The amount of carbon that diffuses out of the martensite lattice, and hence the amount of softer ferrite structure produced, depends on the temperature to which the steel has been heated and the time for which it is held at that temperature. Thus the mechanical properties of the steel can be controlled by the tempering process.

Figure 4.29 shows how, for an oil-quenched steel, the tempering temperature affects the hardness, tensile strength, yield stress and percentage elongation. The higher the tempering temperature the lower the tensile stress, yield stress and hardness but the higher the percentage elongation.

A graph of hardness against tempering temperature is known as

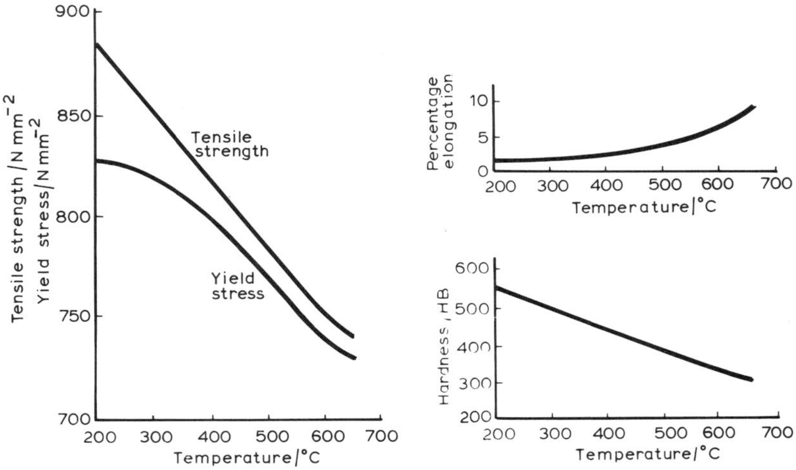

Figure 4.29 The effect of tempering temperature on the properties of a steel (0.40 per cent carbon, 0.70 per cent manganese, 1.8 per cent nickel, 0.80 per cent chromium, 0.25 per cent molybdenum)

the *tempering curve* and often forms part of the data supplied by a steel manufacturer. *Figure 4.30* is a typical example, for a tool steel. With this steel tools for light shock applications where maximum wear resistance is required, e.g. thin sheet punching dies, are tempered at 190°C to 250°C. This gives a hardness of 60 to 63 HRC. Heavy duty applications such as heavy shear blades would be tempered at between 540°C and 560°C and have a hardness of 52 to 56 HRC. For this particular steel the manufacturer recommends that *double tempering* be used. This involves heating to the required tempering temperature, holding at that temperature for at least 60 minutes, then allowing the steel to cool in air before repeating the entire operation.

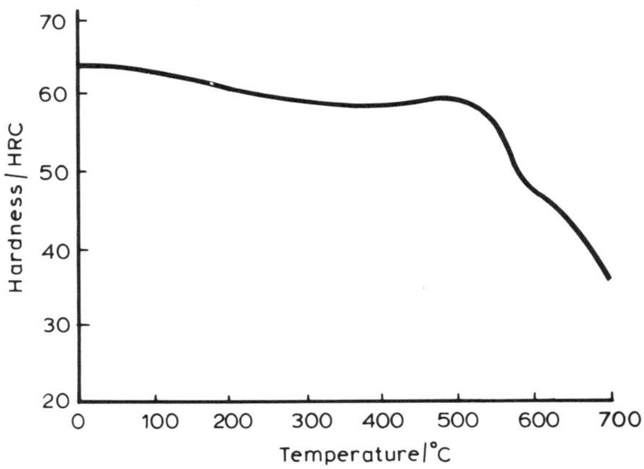

Figure 4.30 Tempering curve for a tool steel K.E.A. 180 (1.55 per cent carbon, 12.0 per cent chromium, 0.85 per cent molybdenum, 0.28 per cent vanadium). (Courtesy of Sanderson Kayser Ltd)

Example
For the tool steel giving the tempering curve in *Figure 4.30,* what tempering temperature should be used if a hardness of 60 HRC is required?

A temperature of about 200°C is indicated by the graph.

ANNEALING

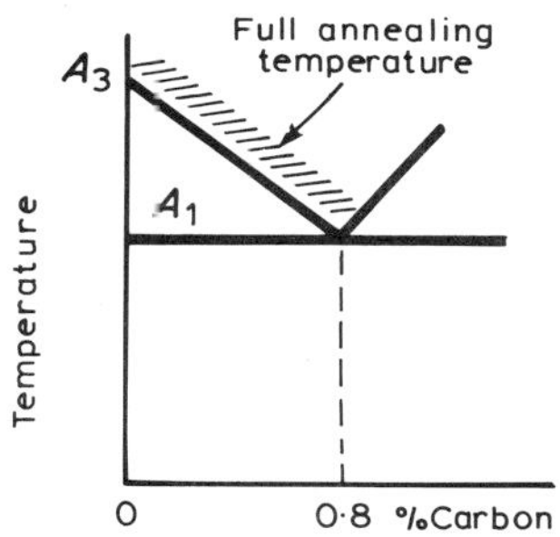

Figure 4.31 The iron-carbon diagram and the temperatures used for full annealing

The term *full annealing* is used for the treatment that involves heating a steel to its austenitic state before subjecting it to very slow cooling. For carbon steels with less than about 0.8% carbon the temperature to which the steel is heated for its microstructure to become austenitic is about 40°C above the A_3 temperature. The result of such a treatment is a microstructure of ferrite and pearlite. This gives a very soft steel. *Figure 4.31* shows the relevant austenising temperature region on the iron–carbon diagram and *Figure 4.32,* the temperature–time relationship for the process with reference to the continuous cooling transformation curve.

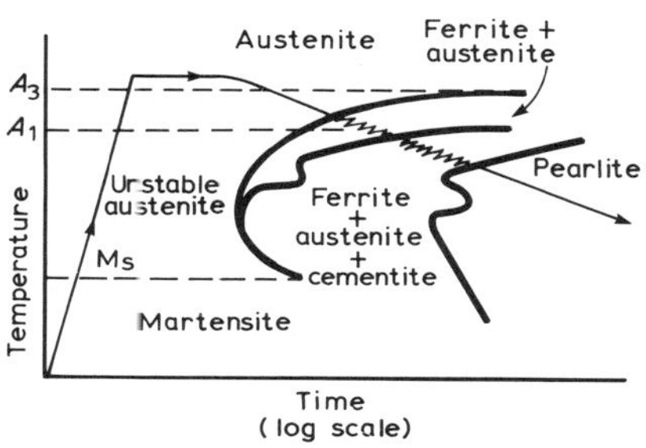

Figure 4.32 The temperature-time relationship for full annealing

For a carbon steel with more than 0.8% carbon, the steel is heated to about 40°C above the A_1 temperature before being slowly cooled. The result of such a treatment is to produce a microstructure of austentite plus cementite at the temperature just above A_1. After cooling, the structure is pearlite plus excess cementite in dispersed spheroidal form. The overall result is a soft steel. The reason for not heating this steel to above the A_3 temperature is that slow cooling of such a steel results in a network of cementite surrounding pearlite. This has the effect of making the steel relatively brittle.

Sub-critical annealing, generally referred to as *process annealing,* is used often during cold-working processes with low-carbon steels, less than about 0.3% carbon, where the material has to be made more ductile for the process to continue. The process involves heating the steel to a temperature just below the A_1 temperature, holding it at that temperature for a while and then allowing the material to cool in air. This is a faster rate of cooling than that employed with full annealing, where the material is cooled in the furnace. This process leads to no change in microstructure, no austenite being produced. The result is however a recrystallisation (see page 29) for a discussion of recrystallisation). Prior to this treatment the crystals may have been deformed by the cold working process, afterwards a new crystal structure occurs with no deformation. The effect of this treatment is to give a reduction in hardness and an increase in percentage elongation. For example, for a 0.15% plain carbon steel:

Condition	*Hardness*/HV	*Percentage elongation*
Cold worked	187	22
Process annealed	162	25

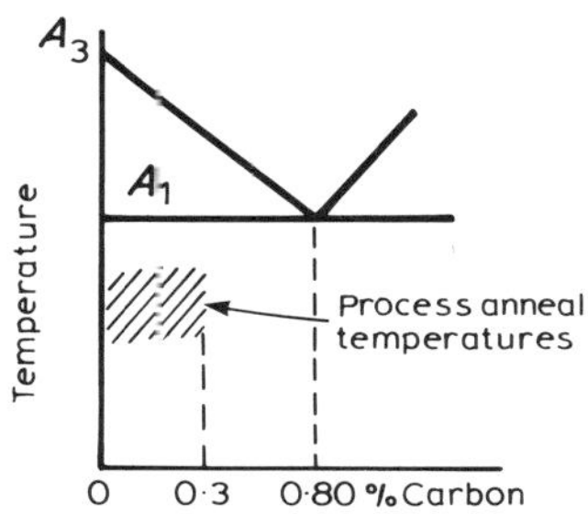

Figure 4.33 The iron-carbon diagram and process annealing

Figure 4.33 shows the region of the iron-carbon diagram relevant to process annealing.

When sub-critical annealing is applied to steels having percentages of carbon higher than 0.3%, the effect of the heating is to cause the cementite to assume spherical shapes. Because of this the process is referred to as *spheroidizing annealing. Figure 4.34* shows the relevant part of the iron–carbon diagram for this treatment. The process results in a reduction in strength and hardness but an increase in ductility and machinability. The spherical pieces of cementite act as chip breakers during machining and so improve machinability.

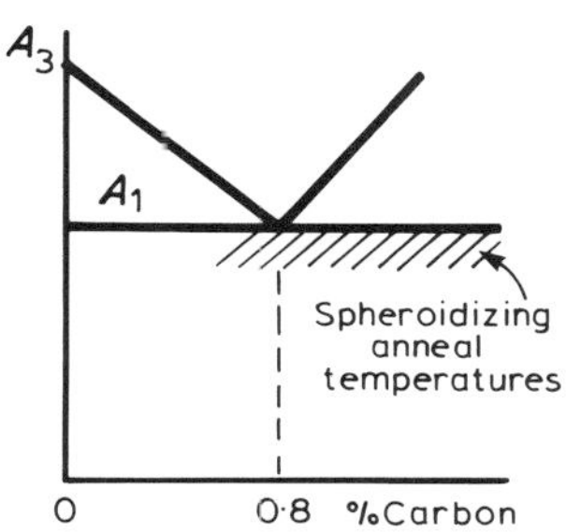

Figure 4.34 The iron-carbon diagram and spheroidizing annealing

The term *normalising* is used to describe an annealing process which involves heating a steel to about 40°C above the A_3 temperature and then allowing the steel to cool in air. The process is thus similar to full annealing in that the microstructure is transformed to austenite but differs in that a faster rate of cooling is

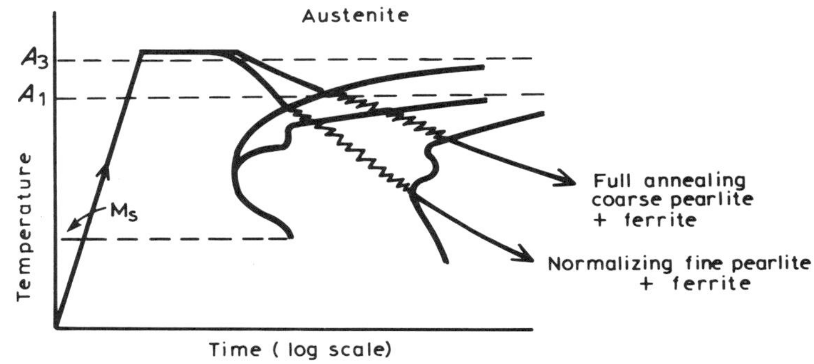

Figure 4.35 The temperature–time relationship for full annealing and normalizing

used (*Figure 4.35*). The result of such a treatment is a ferrite and pearlite microstructure, the hardness and strength being slightly greater than that which would have occurred with full annealing.

AUSTEMPERING

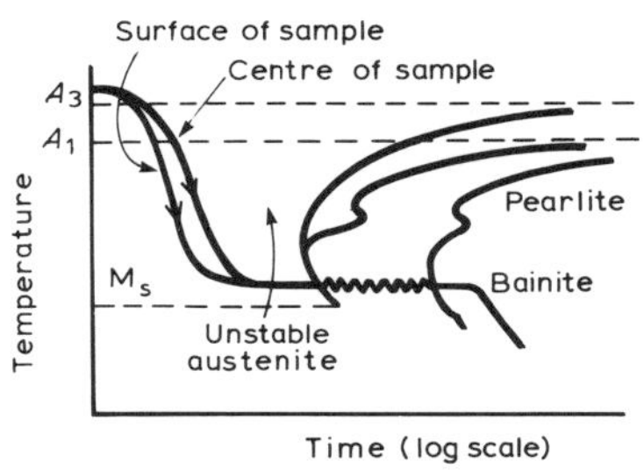

Figure 4.36 Austempering

The *austempering* process involves heating a steel to the austenite state, i.e. above A_3, and then quenching it in a bath held at a temperature above M_s. The steel then remains at that temperature until the austenite is completely transformed into bainite, after which it is allowed to cool. The initial quenching treatment must be at a cooling rate greater than or equal to the critical cooling rate so that no pearlite is formed. The final cooling rate for the bainite structure can be at any rate as no further changes take place. *Figure 4.36* shows the above sequence of events on a TTT diagram.

Bainite structures are softer than martensite structures but have better ductility and impact toughness. Because the steel does not suffer a severe quench treatment it is less likely to crack and distort. The following table shows how the properties of an austempered steel compare with those of the same steel (0.95% carbon steel) when quenched to martensite and then tempered.

Properties	*Austempered*	*Quenched and tempered*
Hardness/HV	545	560
Percentage elongation	11	11
Impact test/J	58	16

MARTEMPERING

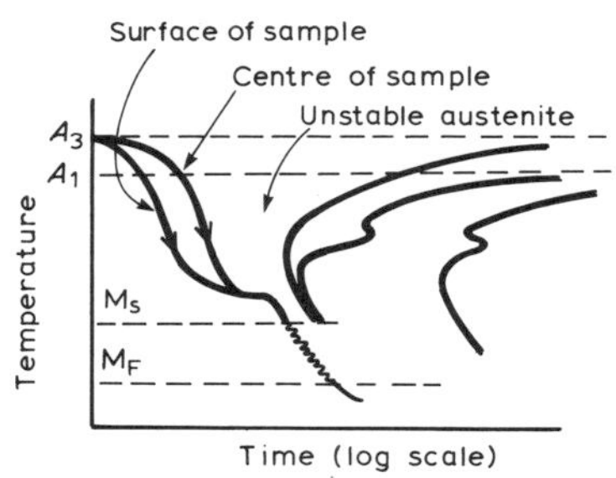

Figure 4.37 Martempering

Martempering is a hardening treatment that consists of heating a steel to its austenitic state, i.e. above A_3, and then quenching it to a temperature just above M_s. The quenching must give a cooling rate faster than the critical cooling rate. The steel is then held at the temperature sufficiently long for the entire piece of steel to come to the same temperature, without any transformation to bainite occurring. It is then cooled in air to change the austenite to martensite. No transformation product other than martensite should occur in this process. *Figure 4.37* shows the above sequence of events on a TTT diagram. After such a treatment the steel is then tempered in the usual way (see page 57).

The effect of such a treatment is to minimise cracking and distortion, the thermal shock of the quenching is reduced. The hardness and ductility are generally similar to those obtained by the direct quenching to the martensitic state followed by tempering. The impact toughness may however be better.

SURFACE HARDENING

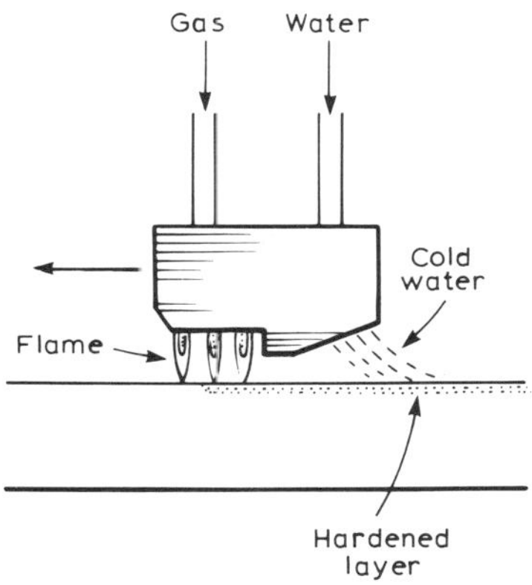

Figure 4.38 A burner with cooling water for flame hardening

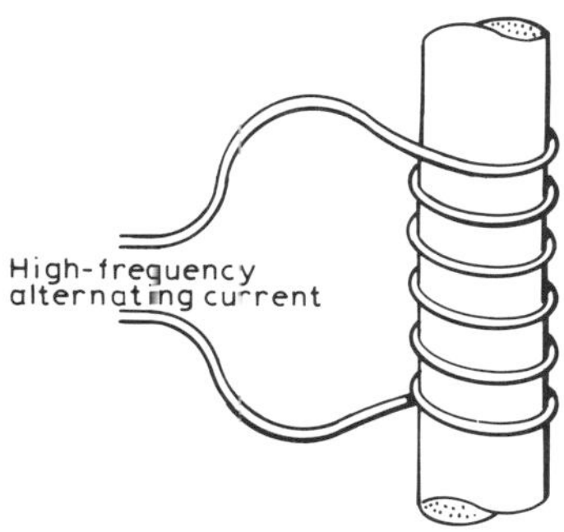

Figure 4.39 Induction hardening of a tube or rod

There are many situations where wear or severe stress conditions may indicate the need for a hard surface to a component without the entire component being made hard and possibly brittle. The various methods used for surface hardening can, in the main, be grouped into two categories:

1. Selective heating of the surface layers;
2. Changing the composition of the surface layers.

One selective heating method is called *flame hardening*. This method involves heating the surface of a steel with an oxy-acetylene flame and then immediately quenching the surface in cold water (*Figure 4.38*). The heating transforms the structure of the surface layers to austenite and the quenching changes this austenite to martensite; the result is a hard layer of martensite in the surface. The depth of hardening depends on the heat supplied per unit surface per unit time. Thus the faster the burner is moved over the surface the less the depth of hardening. This is because less of the steel has its structure transformed into austenite and hence martensite. The temperatures used in the method are typically of the order of 850°C or more, i.e. above the A_3 temperature.

Another method involving selective heating is called *induction hardening*. This method involves placing the steel component within a coil through which a high frequency current is passed (*Figure 4.39*). This alternating current induces another alternating current to flow in the surface layers of the steel component. This induction is called electromagnetic induction. The induced currents heat the surface layers. The temperatures so produced cause the surface layers to change to austenite. The cooling following the passage of the current transforms the austenite to martensite. The depth of heating produced by this method, and hence the depth of hardening, is related to the frequency of the alternating current used. The higher the frequency the less the hardened depth, but the more rapid the temperature rise. The following data illustrates this point:

Frequency used/kHz	*Depth of hardening*/mm
3	4.0 to 5.0
10	3.9 to 4.0
450	0.5 to 1.1

The form of the induction coil used depends on the shape of the component being hardened, also the size of the area to be hardened.

The composition of the steel to be surface hardened by selective heating has to be chosen with certain criteria in mind. Before the surface hardening treatment, the steel must have its inner core with the right mechanical properties. These properties will be unaffected the surface treatment. This means that the composition of the steel and its heat treatment have to be chosen. The heat treatment has also to be chosen with the selective heating process in mind. The selective heating has to be able to change the microstructure of the steel to austenite in a very short amount of time. Hardened and tempered steels respond well but coarse annealed steels do not. In addition the composition of the steel must be such that the quenching part of the

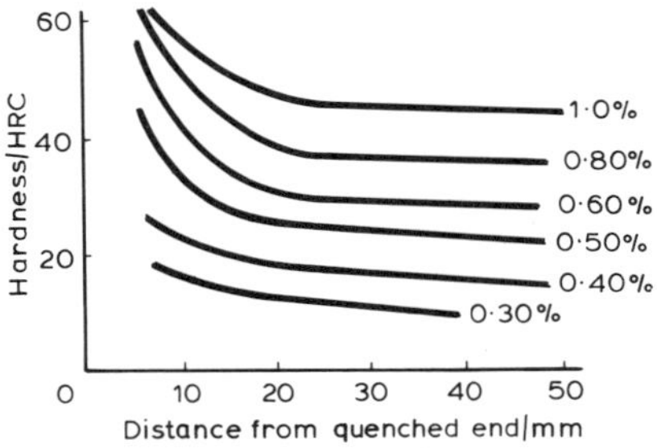

Figure 4.40 Jominy test results showing the effect of percentage carbon on the hardenability of a steel

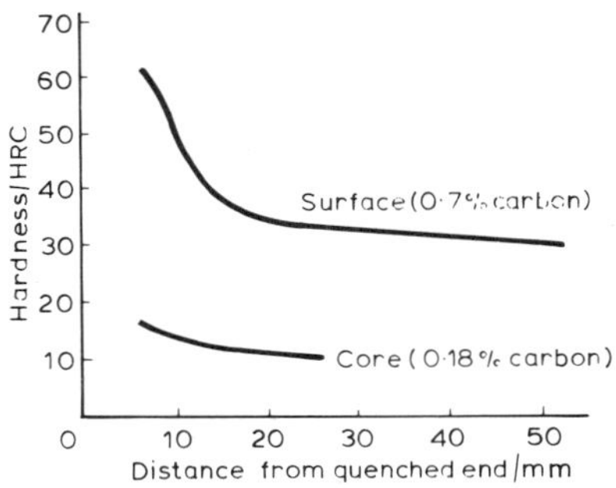

Figure 4.41 Jominy test results for a carburised steel

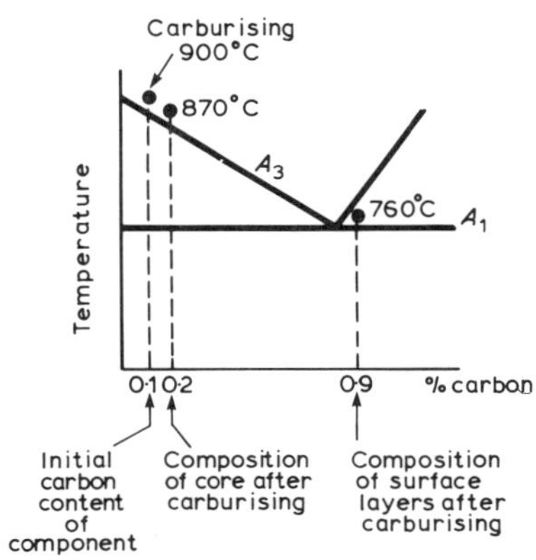

Figure 4.42 Heat treatment for a carburised steel

selective heating process will produce martensite and so harden the steel. This tends to mean carbon contents of 0.4% or more (*Figure 4.40*).

One surface hardening process that involves changing the composition of the surface layers is *carburising*. This method involves increasing the carbon content of the surface layers, followed by a quenching process to convert the surface layers into martensite. This process is normally carried out on a steel containing less than about 0.2% carbon, the carburising treatment being used to give about 0.7 to 0.8% carbon in the surface layers. This wide difference in carbon content is neeeded because the quenching treatment following the carburising will affect both the inner core and the surface layers and it is only the surface layers that are to be hardened, the inner core should remain soft and tough. *Figure 4.41* shows the results of Jominy tests for a 0.18% carbon (0.46% manganese, 1.68% silicon, 0.03% chromium, 0.21% molybdenum) steel for its core where the percentage of carbon remains unchanged and for the surface where the percentage of carbon has been increased to 0.7% by carburising. Thus with a fast rate of cooling the carburised surface may have a hardness of the order of 60 HRC while the inner core is about 15 HRC.

There are a number of carburising methods. With *pack carburising* the steel component is heated to above the A_3 temperature while in a sealed metal box which contains charcoal and barium carbonate. The oxygen present in the box reacts with the carbon to produce carbon monoxide. This carbon-rich atmosphere in contact with the hot steel results in carbon diffusing into the surface austenitic layers. Pack carburising tends to be used mainly for large components or where a thick surface layer has to be hardened.

In *gas carburising* the component is heated to above the A_3 temperature in a furnace in an atmosphere of carbon-rich gas. The result is that carbon diffuses into the surface austenitic layers. Gas carburising is the most widely used method of carburising.

Salt bath carburising, or *cyaniding* as it is often known, involves heating the component in a bath of suitable carbon-rich salts. Sodium cyanide is mainly used. The carbon from the molten salt diffuses into the component. In addition there is also diffusion of some nitrogen into the component. Both carbon and nitrogen can result in a microstructure which can be hardened. The method tends to produce relatively thin hardened layers with high carbon content. This occurs because the carburising takes place very quickly. One of the problems with this method is the health and safety hazard posed by the poisonous cyanide. Another problem is the removal of salt from the hardened component after the treatment. This can be particularly difficult with threaded parts or blind holes.

Carburising may result in the production of a large grain structure due to the time for which the material is held at temperature in the austenitic state. Though the final product may be hard there may be poor impact properties due to the large grain size. A heat treatment might thus be used to refine the grains. A two stage process is required as the carbon content of the surface is significantly different from that of the inner core of the material.

The first stage involves a heat treatment to refine the grains in the core. The component is heated to just above the A_3 temperature for the carbon content of the core (*Figure 4.42*). For a core with 0.2%

carbon this is about 870°C. The component is then quenched in oil. The result is a fine grain core, but the surface layers are rather coarse martensite. The martensite is refined by heating to above the A_1 temperature for the carbon content of the surface layers; if this is 0.9% carbon then the temperature would be about 760°C. The component is then water-quenched. This second stage treatment has little effect on the core but refines the martensite in the outer layers. The treatment may then be followed by a low temperature tempering, e.g. about 150°C, to relieve internal stresses produced by the treatment.

Nitriding involves changing the surface composition of a steel by diffusing nitrogen into it, hard compounds, nitrides, are produced. The process is used with those alloy steels that contain elements that form stable nitrides, e.g. aluminium, chromium, molybdenum, tungsten, vanadium. Prior to the nitriding treatment the steel is hardened and tempered to the properties required of the core. The tempering temperature does however need to be in the region 560°C to 750°C. The reason for this is that the nitriding process requires a temperature up to about 530°C, and this must not be greater than the tempering temperature, as the nitriding process would temper the steel and so change the properties of the core.

Unlike carburising, nitriding is carried out at temperatures below the stable austenitic state. The process consists of heating a component in an atmosphere of ammonia gas and hydrogen, the temperatures being of the order of 500 to 530°C. The time taken for the nitrogen to react with the elements in the surface of the steel is often as much as 100 hours. The depth to which the nitrides are formed in the steel depends on the temperature and the time allowed for the reaction. Even with such long times the depth of hardening is unlikely to exceed about 0.7 mm. After the treatment the component is allowed to cool slowly in the ammonia/hydrogen atmosphere. With most nitriding conditions a thin white layer of iron nitrides is formed on the surface of the component. This layer adversely affects the mechanical properties of the steel, being brittle and generally containing cracks. It is therefore removed by mechanical means or chemical solutions.

Because with nitriding no quenching treatments are involved, cracking and distortion are less likely than with other surface hardening treatments. Very high surface hardnesses can be obtained with special alloys. The hardness is retained at temperatures up to about 500°C, whereas that produced by carburising tends to decrease and the surface become softer at temperatures of the order of 200°C. The capital cost of the plant is however higher than that associated with pack carburising.

Carbonitriding is the name given to the surface hardness process in which both carbon and nitrogen are allowed to diffuse into a steel when it is in the austenitic–ferritic condition. The component is heated in an atmosphere containing carbon and ammonia and the temperatures used are about 800°C to 850°C. The nitrogen inhibits the diffusion of carbon into the steel and, with the temperatures and times used being smaller than with carburising, this leads to relatively shallow hardening. Though this process can be used with any steel that is suitable for carburising it is used generally only for mild steels or low alloy steels.

Ferritic nitrocarburising involves the diffusion of both carbon and

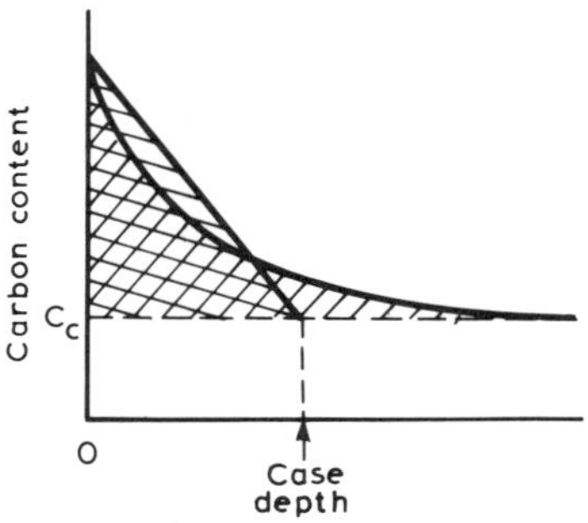

Figure 4.43 Defining case depth

nitrogen into the surface of a steel. The treatment involves a temperature below the A_1 temperature when the steel is in a ferritic condition. A very thin layer of a compound of iron, nitrogen and carbon is produced at the surfaces of the steel. This gives excellent wear and anti-scuffing properties. The process is mainly used on mild steel in the rolled or normalised condition.

The following table compares the main surface hardening treatments. In comparing the processes the term *case depth* is used, which is defined graphically (*Figure 4.43*) in terms of a graph of carbon or nitrogen content against depth under the surface of the steel. A straight line is drawn on the graph such that it passes through the surface carbon content value and so that the area between the line and the core carbon content C_c is the same as that between the actual carbon content graph curve and the core carbon content C_c. Where this line meets the core carbon content line gives the case depth.

Process	*Temperature /°C*	*Case depth /mm*	*Case hardness /HRC*	*Main user*
Pack carburising	810–1100	0.25–3	45–65	Low carbon and carburising alloy steels. Large case depths, large components.
Gas carburising	810–980	0.07–3	45–65	Low carbon and carburising alloy steels. Large numbers of components.
Cyaniding	760–870	0.02–0.7	50–60	Low carbon and light alloy steels. Thin case.
Nitriding	500–530	0.07–0.7	50–70	Alloy steels. Lowest distortion.
Carbo-nitriding	700–900	0.02–0.7	50–60	Low carbon and low alloy steels.
Flame hardening	850–1000	Up to 0.8	55–65	0.4 to 0.7% carbon steels, selective heating.
Induction hardening	850–1000	0.5–5	55–65	0.4 to 0.7% carbon steels, selective heating.

Example

A steel with 0.6% carbon is to be surface hardened. Which type of process, selective heating or carburising, would be suitable?

Carburising is used with steels having less than about 0.2% carbon while selective heating requires steel with 0.4% or more carbon. Thus selective heating is the suitable method.

SELECTING HEAT TREATMENT

Heat treatment can be defined as the controlled heating and cooling of metals for the purpose of altering their mechanical properties. While heat treatment is often used to strengthen or harden a steel there are numerous occasions when heat treatment is used to prepare the material for fabrication, e.g. machining or restoring ductility so that deformation can continue. Full annealing, process annealing, spheroidising annealing and normalising are examples of such treatments. Strengthening and hardening treatments for steels are quenching followed by tempering, austempering and martempering. There is also the possibility of surface hardening, without changing the core properties of the steel concerned.

Suppose a steel has to be softened for some fabrication process. The choice of heat treatment will depend on what hardness is

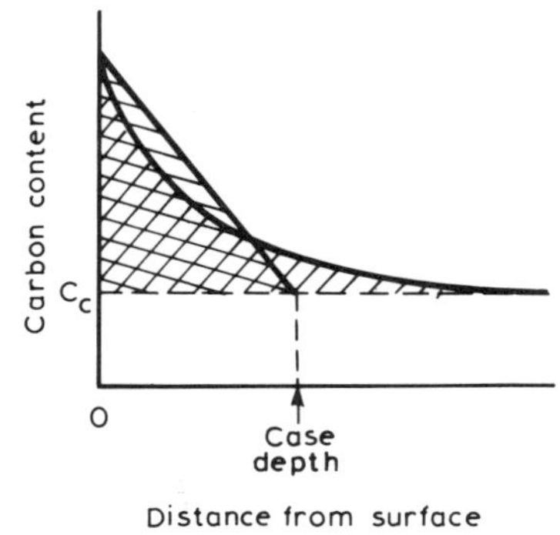

required of the component for the fabrication process. If only a relatively small decrease in hardness and increase in ductility are required and the steel has a low percentage of carbon, then process annealing may be chosen. This would have the advantage of being a faster process than full annealing and so cause less of a production hold-up.

Suppose a steel shaft has to be hardened so that the surface is at a particular hardness and the central core of the shaft has also to be a particular hardness. Consideration has to be given to the hardenability of the steel concerned and whether the required hardness distribution across the shaft section can be obtained by quenching in some particular media followed by tempering.

The selection of a heat treatment depends on the mechanical properties required and the type of steel concerned. Thus in the case of the heat treatment to soften a piece of steel for fabrication, process annealing may be chosen for a low carbon steel but if the steel has more than about 0.3% carbon a different treatment will have to be chosen. Similarly the hardening of the steel shaft depends on the hardenability of the steel concerned. Changing the type of steel used could significantly change the hardenability and hence the method used to produce the required hardness distribution across the shaft section. The size of the shaft also affects the treatment used.

CODING SYSTEMS FOR STEELS

In Great Britain the standard codes for the specification of steels are specified by the *British Standards Institution.* The following codes are those specified to the system given in BS 970 for wrought steels, published between 1970 and 1972. The system uses six symbols.

(1) The first three digits designate the type of steel.

000 to 199	Carbon and carbon-manganese types, the numbers being 100 times the manganese content.
200 to 240	Free cutting steels, the second and third numbers having an approximate relationship to 100 times the mean sulphur content.
250	Silicon-manganese spring steels
300 to 499	Stainless and heat resistant valve steels
500 to 999	Alloy steels

(2) The fourth symbol is a letter.

A The steel is supplied to a chemical composition determined by chemical analysis.
H The steel is supplied to hardenability specification.
M The steel is supplied to mechanical property specification.
S The material is stainless.

(3) The fifth and sixth digits correspond to 100 times the mean percentage carbon content of the steel.

The following examples illustrate the use of the coding.

070M20	A carbon/carbon-manganese type of steel, supplied to mechanical property specification and having 0.20% carbon with 0.70% manganese.
150M19	A carbon/carbon-manganese type of steel supplied to mechanical property specification and having 0.19% carbon with 1.50% manganese.
220M07	A free-cutting steel supplied to mechanical property specification and having 0.07% carbon content. The sulphur content would be about 0.20%
040A04	A carbon/carbon-manganese type of steel supplied to a chemical composition specification and having 0.04% carbon with 0.40% manganese.

Alloy steels have the first three digits in the code between 500 and 999. These are a few examples from this range of steels.

503M40 A nickel steel having 1.0% nickel and 0.40% carbon, being supplied to a mechanical property specification. The number 503 indicates that the steel is a nickel steel.
526M60 A chromium steel with 0.75% chromium and 0.60% carbon, to a mechanical property specification.
530M40 Also a chromium steel, having 1% chromium with 0.40% carbon to a mechanical property specification.
640M40 A nickel-chromium steel with 1.25% nickel, 0.5% chromium and 0.40% carbon, to a mechanical property specification.
653M31 Also a nickel-chromium steel having 3% nickel, 1% chromium and 0.31% carbon, to a mechanical property specification.

A wide range of steels is available. As the specification of a steel is usually according to its tensile strength when in the hardened and tempered condition, a steel can be specified by means of a code letter, the letter indicating the tensile strength range in which the steel falls when in the hardened and tempered condition. The letter is said to refer to the *condition* of the steel.

Condition code	*Tensile strength range*/N mm^{-2}
P	550 to 700
Q	620 to 770
R	700 to 850
S	770 to 930
T	850 to 1000
U	930 to 1080
V	1000 to 1150
W	1080 to 1240
X	1150 to 1300
Y	1240 to 1400
Z	1540 minimum

The following are examples of steels in these various ranges, the range in which a steel occurs depending on not only the composition of the steel but also its size.

Condition	*Grade of steel*	*Limiting ruling section*/mm
P	070M20	19 (0.7% Mn)
P	070M26	29
Q	070M26	13
Q	080M36	29 (0.8% Mn)
R	080M36	13
R	070M55	100 (0.7% Mn)
S	070M55	63
T	070M55	19
R	605M40	150 ($1\frac{1}{2}$% Mn, 0.27% Mo)
S	605M40	100
T	605M40	63
U	605M40	29
T	830M31	250 (3% Ni, 1% Cr, 0.3% Mo)
U	830M31	150
V	830M31	100
W	830M31	63
U	826M40	250 ($1\frac{1}{2}$% Ni, $2\frac{1}{2}$% Cr, $2\frac{1}{2}$% Mo)
V	826M40	250
W	826M40	250
X	826M40	150
Y	826M40	150
Z	826M40	100

The following tables indicate the type of information available for steels when specified to BS 970.

Carbon and carbon-manganese steels : mechanical properties in the hardened and tempered condition related to tensile strength range.

Condition	*P*					*Q*					*R*				
R_m	550–700					620–770					700–850				
HB	152–207					179–229					201–255				
Grade	LRS	R_e	A	I	$R_{p0.2}$	LRS	R_e	A	I	$R_{p0.2}$	LRS	R_e	A	I	$R_{p0.2}$
070M20	19	355	20	40	340										
070M26	29	355	20	40	325	13	415	16	33	400					
080M40						63	385	16	33	355	19	465	16	33	450
150M19	150	340	18	54	310	63	430	16	54	400	29	510	16	40	480
212M36	100	340	20	33	310	63	400	18	33	370	13	495	16	54	480

R_m denotes tensile strength/N mm^{-2}.
HB denotes Brinell Hardness number.
LRS denotes limiting ruling section/mm.
R_e denotes yield stress/N mm^{-2}.
A denotes elongation, minimum on $5.65\sqrt{S_0}$ gauge length %.
I denotes minimum Izod impact value/J.
$R_{p0.2}$ denotes minimum 0.2% proof stress/N mm^{-2}.

Thus if one of the above steels had to be selected to have a hardness of about 200 HB when used in the hardened and tempered condition with a limiting ruling section of 63 mm, the choice would be between 080M40, 150M19 and 212M36. If there was also the condition that the Izod impact value should be not be less than 40 J then the choice would be narrowed to 150M19.

Alloy steels : mechanical properties in the hardened and tempered condition related to tensile strength range.

Condition	*R*					*S*					*T*				
R_m	700–850					770–930					850–1000				
HB	201–255					223–277					248–302				
Grade	LRS	R_e	A	I	$R_{p0.2}$	LRS	R_e	A	I	$R_{p0.2}$	LRS	R_e	A	I	$R_{p0.2}$
503M40	63	525	17	47	510	22	585	15	27	570					
530M60	100	525	17	54	510	63	585	15	54	570	29	680	13	54	665
605M36	250	495	15	33	480	100	585	15	54	570	63	680	13	54	665
	150	525	17	54	510										
640M40	150	525	17	54	510	100	585	15	54	570	63	680	13	54	665
653M31						150	585	15	40	570	100	680	13	54	665
709M40	250	495	15	33	480	250	555	13	27	540					
						150	585	15	54	570	100	680	13	54	665
817M40											250	650	13	41	635
											150	680	13	54	665

The symbols used above are the same as for the previous table.

If there is a requirement for an alloy steel from the above list to meet a specification of yield stress 585 N mm^{-2} and a limiting ruling section of 100 mm, then possible choices would be 605M36 and 640M40. Suppose a steel from the above list was to meet a specification of tensile strength in the range 850-1000 N mm^{-2} and a limiting ruling section of 250 mm, then the choice would be 817M40. With this size of ruling section the Izod impact value would be 41 J.

Other countries have other code systems for specifying steels. A common one is that developed by the *American Society of Automotive Engineers,* the code being referred to as *SAE* numbering. A four-symbol code is used, the first two numbers indicating the type of steel and the third and fourth numbers indicating 100 times the percentage of carbon content.

1000 series	Carbon steels; 10XX is plain carbon with a maximum manganese content of 1.00%, 11XX is resulphurised, 12XX resulphurised and rephosphorised, 13XX has 1.75% manganese, etc.
2000 series	Nickel steels; 23XX are nickel steels having 3.50% nickel, 25XX has 5.00% nickel.

3000 series	Nickel-chromium steels; 32XX has 1.75% nickel and 1.07% chromium, 34XX has 3.00% nickel and 0.77% chromium.
4000 series	Chromium-molybdenum steels; 40XX has 0.20 or 0.25% molybdenum, 41XX has 0.50%, 0.80% or 0.95% chromium with 0.12%, 0.20%, 0.25% or 0.30% molybdenum.
5000 series	Chromium steels containing up to 1% chromium.
6000 series	Chromium-vanadium steels.
7000 series	Steels containing up to 6% tungsten.
8000 series	Steels containing nickel, chromium and molybdenum.
9000 series	Silicon-manganese steels.

There are some specially specified steels which have modified SAE numbers.

XXBXX	The 'B' denotes boron intensified steels, the letter being incorporated in the middle of the normal specifications.
XXLXX	The 'L' denoted leaded steels, the letter being incorporated in the middle of the normal specification.

These examples illustrate the use of the coding.

1330	A carbon steel having 0.30% carbon and 1.75% manganese.
4012	A steel having 0.12% carbon, 0.20% molybdenum and 0.90% manganese.

Tool steels are specified by BS 4659, the coding used being based on that of the American Iron and Steel Institute (AISI), the only difference being that in Britain the American codes are prefixed by a B:

BS 4659 code	*AISI code*	
BW	W	A water-hardening tool steel.
BO	O	Oil-hardening tool steel for cold work.
BA	A	Medium alloy hardening for cold work.
BD	D	High-carbon and high-chromium content for cold work.
BH	H	Chromium or tungsten base for hot work.
BT	T	Tungsten base, high-speed steel.
BM	M	Molybdenum base, high-speed steel.
BL	L	Low alloy tool steel for special applications.

The tool steel code letters are followed by a number to denote a particular steel composition. Thus BW5 is a water-hardening tool steel having 1.10% carbon and 0.50% chromium. BO7 is an oil-hardening tool steel having 1.20% carbon, 1.75% tungsten and 0.75% chromium.

PROBLEMS

1. Explain the effect on the mechanical properties of plain carbon steels of an increase in carbon content.

2. In what ways do alloying elements, other than carbon, have an effect on the properties of steel?

3. In what main ways do the following elements affect the properties of steel when they are alloyed with it?
(a) Manganese, (b) Chromium, (c) Molybdenum, (d) Sulphur, (e) Silicon.

4. What elements added to a steel can improve the corrosion properties of that steel?

5. What elements are usually added to a steel to increase its machinability?

6. State two elements that, when added to steel, restrict grain growth.

7. What is meant by austenitic steels and ferritic steels?

8. A commonly used structural steel has the following composition: 0.40% carbon, 0.55% manganese, 1.50% nickel, 1.20% chromium, 0.30% molybdenum. What are the effects of the nickel, chromium and molybdenum on the properties of the steel?

9. What are the properties required of (a) tool steels in general, (b) tool steels for use in hot working processes, (c) shock-resistant tool steels?

10. What are stainless steels?

11. What is meant by 'free-cutting' steels?

12. How can (a) ferritic steels, (b) martensitic steels be hardened?

13. Explain what is meant by the term critical points and, in this context, the symbols A_1, A_3 and A_{cm}.

14. Why is the iron carbon diagram, as in *Figure 4.2,* called an equilibrium diagram?

15. What is shown by a TTT diagram?

16. Explain how a TTT diagram is obtained.

17. Sketch typical TTT diagrams for plain carbon steels having (a) the eutectoid composition, (b) less carbon than the eutectoid composition.

18. What is the effect on the TTT diagram of a carbon steel of the addition of chromium to the alloy?

19. What are continuous cooling transformation curves?

20. What is meant by the critical cooling rate? How is the critical cooling rate determined by the continuous cooling transformation curve?

21. Explain, with the aid of a continuous cooling transformation curve, how the structures produced by cooling a plain carbon steel are determined by the rate of cooling. Consider cooling rates both greater and less than the critical cooling rate.

22. Why must water quenching be used to harden plain carbon steels but some alloy steels can be hardened by air cooling?

23. Explain the term hardenability.

24. Describe the Jominy test.

25. Explain how Jominy test results can be used to determine the hardness distribution that would occur across a section of a steel component when it is quenched.

26. Use *Figure 4.27* for this question. (a) What would be the hardness at the centre of a 12 mm diameter bar of 0.40% plain carbon steel when it is quenched in still water? (b) What would the answer to (a) have been if still oil had been used for the quenching? (c) For the alloy steel given in the figure, what would be the maximum diameter of bar that could be quenched in still water if the hardness at the centre of the bar had to be 600 HV or more?

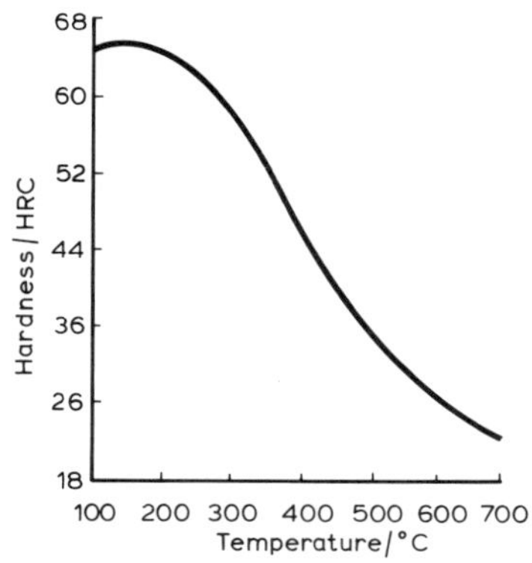

Figure 4.44 Tempering curve for a 1.4 per cent plain carbon steel

27. Explain, using a continuous cooling transformation curve, why the structure at the centre of a bar may differ from that at the surface.

28. Explain the term 'limiting ruling section.'

29. Explain what the tempering process is and why the mechanical properties of a tempered steel depend on the tempering temperature.

30. With the tempering curve given in *Figure 4.30* for the K.E.A. 180 steel, what tempering temperature should be used if the steel is to have a hardness of 45 HRC?

31. *Figure 4.44* shows the tempering curve for a 1.4% plain carbon steel. What would be the hardness of such a steel after tempering at 300°C?

32. The tempering curve in *Figure 4.44* was for a plain carbon steel that had been heated to 1200°C, water quenched and then cooled to a subzero temperature. (a) What was the purpose of the heating to 1200°C? (b) Why was the steel cooled to below zero?

33. Describe the following heat treatment processes. (a) Full annealing, (b) Process annealing, (c) Spheroidising annealing, (d) Normalising.

34. In what ways do the processes listed in question 33 differ?

35. Explain, using a TTT diagram, how full annealing and normalising can lead to different microstructures.

35. Explain using TTT diagrams the (a) austempering and (b) martempering processes.

36. Explain how flame hardening leads to the surface hardening of a steel.

37. Explain, by means of *Figure 4.40,* how increasing the carbon content in the surface layers of a carbon steel leads to an increase in its hardness.

38. Compare the surface hardening processes of nitriding and carburising.

39. Which surface hardening treatment would be most appropriate for an alloy steel which cannot be heated above 550°C?

40. What factors have to be considered in determining the type of quenching to be used to harden a steel?

41. State the type of steel and the percentages of the various constituents for the steels specified by the following BS 970 codes. (a) 040A12, (b) 070M26, (c) 150M19, (d) 210M15, (e) 503A37, (f) 653M31.

42. A steel is specified as having the condition 'R' when at a certain ruling section. What is meant the condition 'R' and why is the ruling section specified?

43. Tool steel is specified as being 'BD' according to BS 4659. What is the significance of the letters BD?

44. A chromium-base tool steel is required for hot work. What code letters should be looked for in tool steels quoted to BS 4659 specifications?

45. A steel specified to SAE standards is quoted as being 1020. What is the significance of the 1 in this number?

46. A steel is said to be in the SAE 3000 series. What does this signify with regard to the composition of the steel?

47. What would be the codes used for the following steels in terms of BS 970?
(a) A plain carbon steel with 0.12% carbon and 0.40% manganese specified to mechanical properties. (b) A plain carbon steel with 0.78% carbon and 0.70% manganese to chemical composition specification. (c) A free cutting steel with 0.36% carbon and 2.12% sulphur specified to mechanical properties.

48. Select steels from the table of carbon and carbon–manganese steels to BS 970 given on page 67 to meet the following specifications.
(a) Hardness 152-207 HB, limiting ruling section 150 mm. (b) Tensile strength 620-770 N mm^{-2}, limiting ruling section 63 mm, Izod impact value 54 J. (c) Tensile strength 700-850 N mm^{-2}, elongation at least 16%, limiting ruling section 29 mm.

49. Select steels from the table of alloy steels to BS 970 given on page 67 to meet the following specifications.
(a) Condition R, limiting ruling section 150 mm, Izod impact value at least 50 J. (b) Tensile strength 850-1000 N mm^{-2}, limiting ruling section 100 mm. (c) Hardness 250 HB, limiting ruling section 100 mm, tensile strength 900 N mm^{-2}.

50. Select the quenching medium to be used in the following instances. Use *Figure 4.15* and *Figure 4.16* as sources of information. They refer to steels 070M38 and 817M40. (a) Steel 070M38, 10 mm diameter bar to be entirely martensitic. (b) What is the maximum diameter bar of 070M38 that can be water quenched and give a martensitic structure? (c) Steel 070M38, 20 mm diameter bar to be entirely bainite. (d) Steel 817M40, 50 mm diameter bar to be entirely martensitic. (e) What is the maximum diameter bar of 817M40 that can be air cooled and give a martensitic structure?

5 Polymers

Objectives: At the end of this chapter you should be able to:
Explain how polymer crystallinity affects polymer properties.
Describe the type of polymer molecular chain that leads to crystallinity.
Explain the significance of the glass transition temperature.
Explain the effects of temperature and time on the properties of a polymer.
Explain the effect of orientating polymer chains on the properties of the polymer.
Explain what is meant by internal and external plasticisation.
Compare the mechanical properties of commonly used thermoplastics, thermosets and elastomers.

STRUCTURE OF POLYMERS

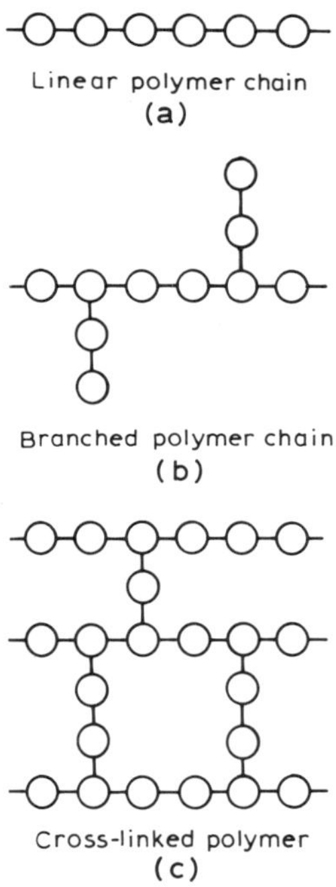

Figure 5.1 (a) Linear polymer chain, (b) Branched polymer chain, (c) Cross-linked polymer

Polymers involve the combining of many small molecules to give a large molecule involving very large numbers of atoms. The resulting large molecule may be in the form of a long linear chain, a chain with side branches or a cross-linked polymer giving rise to a network of interlocked atoms (*Figure 5.1*).

A crystal can be considered to be an orderly packing-together of atoms. The molecular chains of a polymer may however be completely tangled up in a solid with no order whatsoever (*Figure 5.2*). Such a material is said to be *amorphous.* There is however the possibility that the polymer molecules can be arranged in an orderly manner within a solid. Thus *Figure 5.3* shows linear polymer molecules folded to give regions of order. The orderly parts of such polymeric materials are said to be *crystalline.* Because long molecules can easily become tangled up with each other, polymeric materials are often only partially crystalline, i.e. parts of the material have orderly arrangements of molecules while other parts are disorderly.

Not all polymers can give rise to crystallinity. It is most likely to occur with simple linear chain molecules. Branched polymer chains are not easy to pack together in a regular manner, the branches get in the way. If the branches are completely regularly spaced along the chain then some crystallinity is possible, irregularly spaced branches render crystallinity improbable. Cross-linked polymers cannot be rearranged due to the links between chains and so crystallinity is not possible.

There are many polymers based on the form of the polyethylene molecule. Despite being linear molecules they do not always give rise to crystalline structures. PVC is essentially just the polyethylene molecule with some of the hydrogen atoms replaced with chlorine atoms (*Figure 5.4*). The molecule does not however give rise to crystalline structures. This is because the chlorine atoms are rather bulky and are not regularly spaced along the molecular chain. It is this lack of regularity which makes packing of the PVC molecular chains too difficult. Polypropylene has a molecule rather like that of

Figure 5.2 A linear amorphous polymer. Individual atoms are not shown, the chains just being represented by lines

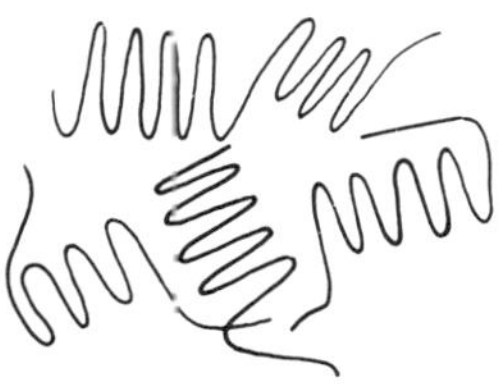

Figure 5.3 Folded linear polymer chains

polythene but with some of the hydrogen atoms replaced with CH_3 groups. These are however regularly spaced along the molecular chain and thus orderly packing is possible and so some degree of crystallinity.

The following table shows the form of the molecular chains and the degree of crystallinity possible for some common polymers.

Polymer	*Form of chain*	*Possible crystallinity*
Polyethylene	Linear	95%
	Branched	60%
Polypropylene	Regularly spaced side groups on linear chain	60%
Polyvinyl chloride	Irregularly spaced bulky chlorine atoms (fig. 5.4)	0
Polystyrene	Irregularly spaced bulky side groups	0

When an amorphous polymer is heated it shows no definite melting temperature but progressively becomes less rigid. The molecular arrangement in an amorphous material is all disorderly, just like that which occurs in a liquid. It is for this reason, i.e. no structural change occurring, that no sharp melting point occurs. For crystalline polymers there is an abrupt change at a particular temperature. Thus if the density of the polymer were being measured as a function of temperature, an abrupt change in density would occur at a particular temperature. At this temperature the crystallinity of the polymer disappears, the structure changing from a relatively orderly one below the temperature to a disorderly one above it. The temperature at which the crystallinity disappears is defined as being the *melting point* of the polymer. The following are the melting points of some common polymers.

Polymer	*Crystallinity*	*Melting point/°C*
Polyethylene	95%	138
Polyethylene	60%	115
Polypropylene	60%	176
Polyvinyl chloride	0%	212*
Polystyrene	0%	—

* This is not a clear melting point but a noticeable softening region of temperature.

The degree of crystallinity of a polymer affects its mechanical properties. The more crystalline a polymer the higher its tensile modulus. Thus the linear chain form of polyethylene in its crystalline form has the molecules closely packed together. Greater forces of attraction can exist between the chains when they are closely packed.

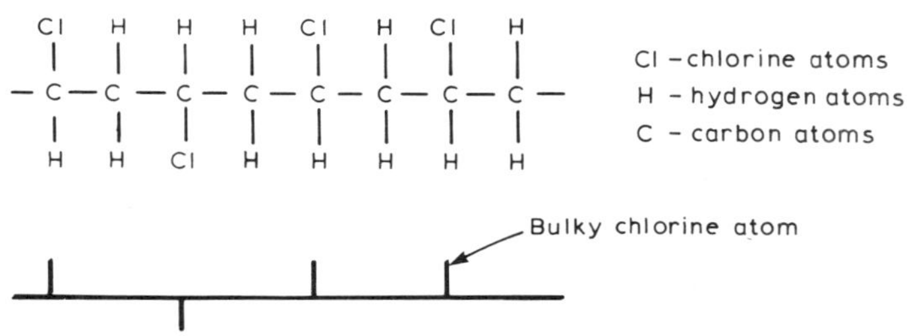

Figure 5.4 The basic form of a p.v.c. molecule. The chlorine atoms are generally irregularly arranged on the different sides of the chain, so rendering orderly packing of the chains difficult

The result is a stiffer material, a material with a higher tensile modulus. The branched form of polyethylene has a lower crystallinity and thus a lower tensile modulus. This is because the lower degree of crystallinity means that the molecules are not so closely packed together, an orderly structure is easier to pack closely than a disorderly one (you can get more clothes in a drawer if you pack them in an orderly manner than if you just throw them in). The further apart the molecular chains the lower the forces of attraction between them and so the less stiff, i.e. more flexible, the material and hence the lower the tensile modulus. The following are the tensile modulus and tensile strength values for polyethylene with different degrees of crystallinity.

Polymer	*Crystallinity*	*Tensile modulus* /kN mm^{-2}	*Tensile strength* /N mm^{-2}
Polyethylene	95%	21 to 38	0.4 to 1.3
Polyethylene	60%	7 to 16	0.1 to 0.3

The more cross-links there are with a polymer the stiffer the polymer is, i.e. the higher its tensile modulus. A highly cross-linked polymer may be a rather hard, brittle substance.

GLASS TRANSITION TEMPERATURE

P.V.C., without any additives and at room temperature, is a rather rigid material. It is often used in place of glass. But if it is heated to a temperature of about 87°C a change occurs, the p.v.c. becomes flexible and rubbery. The p.v.c. below this temperature gives only a moderate elongation before breaking, above this temperature it stretches a considerable amount and behaves rather like a strip of rubber.

Polythene is a flexible material at room temperature and will give considerable extensions before breaking. If, however, the polythene is cooled to below about −120°C it becomes a rigid material.

The temperature at which a polymer changes from a rigid to a flexible material is called the *glass transition temperature.* The material is considered to be changing from a glass-like material to a rubber-like material. The following are the glass-transition temperatures for some common thermoplastics.

Material	*Glass transition temperature/°C*
Polyethylene	−120
Polypropylene	−10
Polyvinyl chloride	87
Polystyrene	100

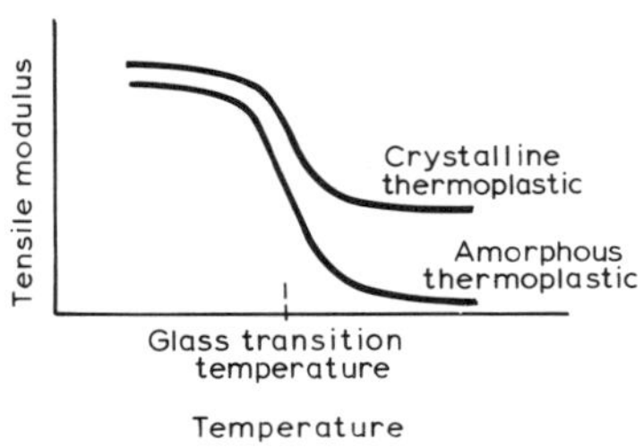

Figure 5.5 The effect of temperature on the tensile modulus

At the glass transition temperature the tensile modulus shows an abrupt change from a high value for the glass-like material to a low value for the rubber-like material. *Figure 5.6* shows the form of the stress-strain graph for a polymer both below and above this transition temperature. Below the transition temperature the material has the type of stress–strain graph characteristic of a relatively brittle substance; above the transition temperature the graph is more like that of a rubber-like material in that the polymer may be stretched to many times its original length.

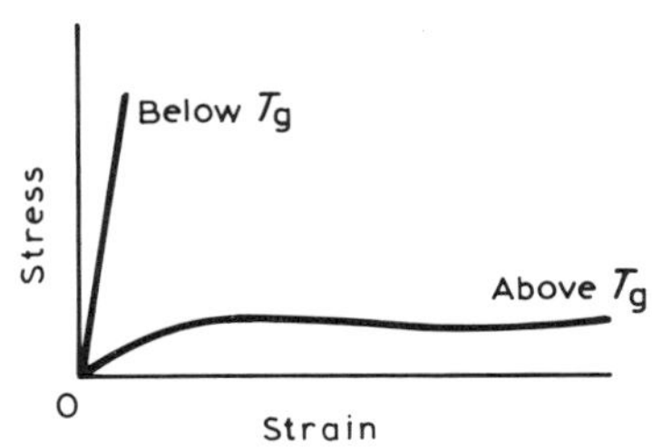

Figure 5.6 Stress-strain graphs for a polymer below and above its glass transition temperature T_g

Above the glass transition temperature the polymer chains are reasonably free to move and thus the application of stress causes molecular chains to uncoil and/or slide over one another, hence producing large extensions. When the temperature is decreased the density of the polymer increases and the chains become more closely packed. At the glass transition temperature the packing has become close enough to hinder the movement of the polymer chains to such an extent that the application of stress results in little extension.

The value of the glass transition temperature varies from one polymer to another. It is lowest for those polymers that have linear molecules, e.g. polythene, which are flexible and easily packed together in the solid. Polymers having branched chains have higher glass transition temperatures, e.g. polyvinyl chloride, and are thus not easily packed, if at all, in an orderly manner.

Example
Should a polymer be below or above its glass transition temperature if it is to have (a) a high value of tensile modulus, (b) high elongations?

(a) Below the glass transition temperature (see *Figure 5.5*).
(b) Above the glass transition temperature, as it then behaves more like a rubber. Below the glass transition temperature the material is relatively brittle and thus has a low elongations.

THE EFFECTS OF TEMPERATURE AND TIME ON THE MECHANICAL PROPERTIES

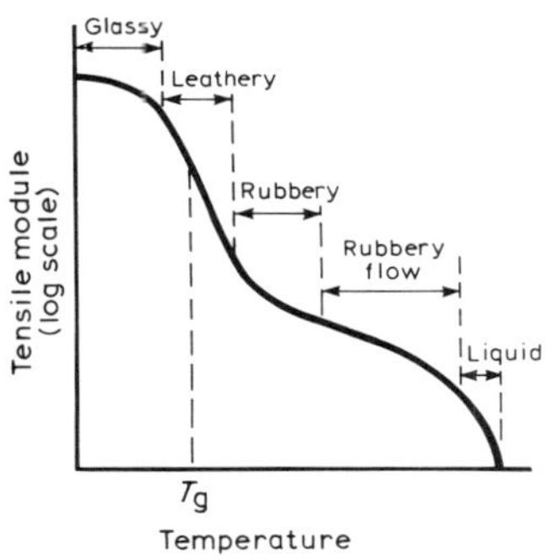

Figure 5.7 Variation of tensile modulus with temperature for an amorphous polymer

For polymers the variation of the tensile modulus with temperature is of considerable significance in that, for instance, it determines whether the plastic spoon used to stir the coffee is stiff and useful for the job concerned or far from stiff and not particularly useful. The tensile modulus of polymers decreases with an increase in temperature. The overall way in which the modulus changes with temperature is the same for all non-crystalline polymers, the differences between polymers being just the actual temperatures at which the changes occur. *Figure 5.7* shows the general type of variation that occurs with temperature — note the log scale for the modulus axis of the graph.

In the glassy range of the graph the tensile modulus is at its maximum value and is less dependent on temperature than at other temperatures. In this region the polymer is stiff and not so easily stretched. This is because the polymer chains are too closely packed to slide past each other and the only way the polymer can extend is by stretching the chains themselves.

The region between the glassy region and the rubbery region is referred to as the leathery region. This represents the transition of the polymer from a tightly packed structure where no chain movement was possible to a state where the polymer chains can uncoil from their tangled state, but spring back to their tangled state when the applied stress is removed.

At yet higher temperatures the chains begin to slip past each other when stress is applied. This is the rubbery flow region. At yet higher temperatures the polymer becomes liquid and complete movement of polymer chains past each other occurs. When chains move past each other the deformation is irrecoverable.

A plastic coffee spoon at room temperature is fairly rigid and if stressed may break in a brittle manner. The spoon is in the glassy region. If the spoon is in boiling water it behaves completely differently, very little stress has to be applied to cause a permanent irrecoverable deformation due to chain slipping occurring, the rubbery flow region.

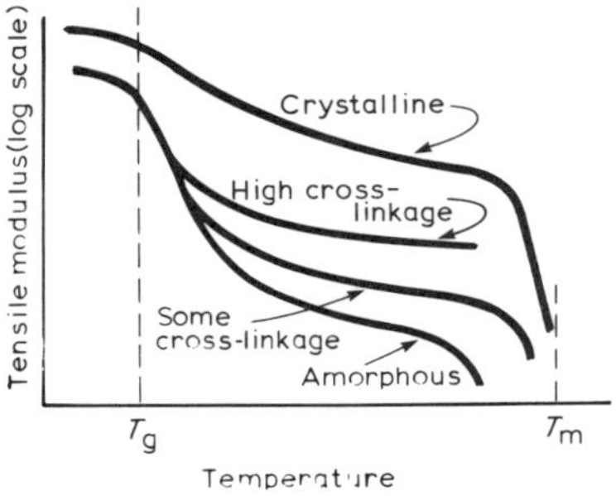

Figure 5.8 Variation of tensile modulus with temperature for polymers

The above discussion has considered what happens when stress is applied to polymers, but no mention was made of whether the stress was applied quickly or slowly. Time is needed for uncoiling or movement of polymer chains. A polymer that, at some particular temperature, may be rubbery with a slow application of stress, may be quite glass-like with a faster application of stress. The effect of increasing the rate of application of stress is to make the polymer more brittle and have a higher tensile modulus.

The effect of crystallinity on the behaviour of a polymer when subject to stress is to make it more stiff, i.e. have a higher tensile modulus. The polymer also becomes more brittle. *Figure 5.8* shows how the tensile modulus varies with temperature for a crystalline polymer.

The graph also shows how cross-links in a polymer affect its properties. With an increasing amount of cross-linkage so the rubbery and flow ranges disappear. With a high amount of cross-linkage the material is hard and fairly brittle.

TEMPERATURE AND POLYMER USE

Amorphous polymers tend to be used below their glass transition temperature T_g. They are however formed and shaped at temperatures above the glass transition temperature when they are in a soft condition. Crystalline polymers are however used up to their melting temperature T_m. They can be hot-formed and shaped at temperatures above T_m, being cold-formed and shaped at temperatures between T_g and T_m.

Polythene is a crystalline polymer, with a melting point of 138°C and a glass transition temperature of −120°C for the form that gives 95% crystallinity. The maximum service temperature of polythene items is about 125°C, i.e. just below the melting point. The form of polythene with 60% crystallinity has a melting point of 115°C and a maximum service temperature of about 95°C.

Polystyrene is an amorphous polymer with a glass transition temperature of 100°C. It has a maximum service temperature of about 80°C, i.e. just below the glass transition temperature.

Example
What would you expect to be the maximum service temperature for polyvinyl chloride? It is an amorphous polymer with a glass transition temperature of 87°C.

An amorphous polymer is used below its glass transition temperature and thus a maximum service temperature of about 70°C can be expected.

ORIENTATION

Figure 5.9 shows a typical stress–strain graph for a crystalline polymer, e.g. polythene. When the stress reaches point 'A' the material shows a sudden large reduction in cross-sectional area at some point (*Figure 5.10*). After this initial necking a considerable increase in strain takes place at essentially a constant stress, as the necked area gradually spreads along the entire length of the material. When the entire piece of material has reached the necked stage an increase in stress is needed to increase the strain further.

The above sequence of events can be explained by considering the orientation of the polymer chains. Prior to necking starting, the

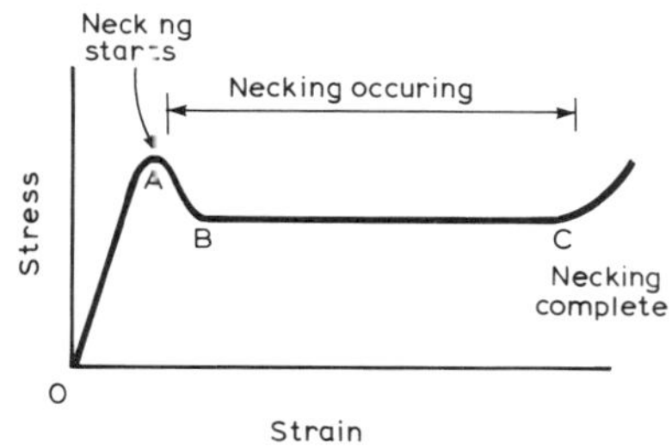

Figure 5.9 Typical stress–strain graph for a crystalline polymer

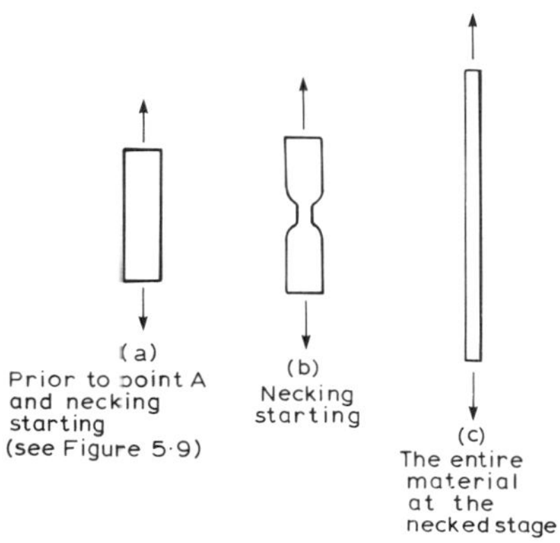

Figure 5.10 Necking with a polymer, (a) Prior to point 'A' and necking starting, (b) Necking starting, (c) The entire material at the necked stage

polymer chains are folded to give regions of order in the material (as in *Figure 5.3*). When necking starts the polymer chains unfold to give a material with the chains lying along the direction of the forces stretching the material. As the necking spreads along the material so more of the chains unfold and line up. Eventually when the entire material is at the necked stage all the chains have lined up. The material in this state behaves differently to earlier in the stress–strain graph. The material is said to be *cold drawn.* It is completely crystalline, i.e. all the chains are packed in a very orderly manner.

The above sequence of events only tends to occur if the material is stretched slowly and sufficient time elapses for the molecular chains to slide past each other. If a high strain rate is used the material is likely to break without it becoming completely orientated. You can try pulling a strip of polythene for yourself and see the changes. (A strip cut from a polythene bag can be used and you can pull it in your hands.)

With crystalline or semi-crystalline polymers orientation is produced at temperatures both below and above the glass transition temperature. With amorphous polymers orientation is produced at temperatures above the glass transition temperature. Below that temperature an amorphous polymer is too brittle and breaks. The effect of the chains becoming orientated is to give a harder, stronger material. The effect can be considered to be similar to work hardening with metals.

In order to improve the strength of polymer fibres, e.g. polyester fibres, they are put through a drawing operation to orientate the polymer chains. Stretching a polymer film causes orientation of the polymer chains in the direction of the stretching forces. The result is an increase in strength and stiffness in the stretching direction. Such stretching is referred to as uniaxial stretching and the effect as *uniaxial orientation.* The material is however weak and has a tendency to split if forces are applied in directions other than those of the stretching forces. For the polyester fibres this does not matter as the forces will be applied along the length of the fibre, with the film this could be a serious defect. The problem can be overcome by using a biaxial stretching process in which the film is stretched in two directions at right angles to each other. The film has then *biaxial orientation.*

Rolling through compression rollers is similar in effect to the drawing operation and results in an uniaxial orientated product. Extrusion has a similar effect.

Orientation can be obtained by both cold- and hot-working processes. Hot working involves working at temperatures just below the glass transition temperature. Under such conditions orientation can be produced without any internal stresses being developed. Cold working, at lower temperatures than those of hot working, requires more energy to produce orientation but also results in internal stresses being produced.

If orientated polymers are heated to above their glass transition temperature they lose their orientation. On cooling they are no longer orientated and are in the same state as they were before the orientation process occurred. This effect is made use of with shrinkable films. The polymer film is stretched, and so made longer even when the stretching force is removed. If it then wrapped around some package and heated, the film contracts back to its initial, prestretched state. The result is a plastic film tightly fitting the package. The film is said to show *elastic memory.*

PLASTICISATION

The term *plastic* is used to describe materials based on polymers. Such materials invariably contain other substances which are added to the polymers to give required properties. *Stabilisers* are substances added in order to protect the plastic against ultraviolet radiation. *Fillers* may be added to change the strength properties, e.g. the addition of glass fibres, or reduce friction, e.g. graphite. An important group of additives are called *plasticisers.* Their primary purpose is to enable the molecular polymer chains to slide more easily past each other.

Internal plasticisation involves modifying the polymer chain by the introduction into the chain of bulky side groups. An example of this is the plasticisation of polyvinyl chloride by the inclusion of some 15% of vinyl acetate in the polymer chains. These bulky side groups have the effect of forcing the polymer chains further apart, so reducing the attractive forces between the chains and so permitting easier flow of chains past each other.

A more common method of plasticisation used is called *external plasticisation.* It involves a plasticiser being added to the polymer, after the chains have been produced. This plasticiser may be a liquid which disperses throughout the plastic, filling up the spaces between the chains. The effect of the liquid is the same as adding a lubricant between two metal surfaces, the polymer chains slide more easily past each other. The effect of the plasticiser is to weaken the attractive forces that exist between the polymer chains. The plasticiser decreases the crystallinity of polymers as it tends to hinder the formation of orderly arrays of polymer chains. The plasticiser also reduces the glass transition temperature. The effect on the mechanical properties is to reduce the tensile strength and increase the flexibility. The following table shows the effects of plasticiser on the mechanical properties of polyvinyl chloride.

	Tensile strength /N mm^{-2}	*Elongation* %
No plasticiser	52 to 58	2 to 40
Low plasticiser	28 to 42	200 to 250
High plasticiser	14 to 21	350 to 450

The p.v.c. with no plasticiser is a rigid material, with low plasticiser content the p.v.c. is pliable, with very high plasticiser content the p.v.c. is soft and rubbery.

PROPERTIES OF COMMON THERMOPLASTICS

The following are brief outlines of the main properties and uses of thermoplastics commonly used in engineering (*Figures 5.11* and *5.12*):

Polyethylene (polythene) is referred to as high density when the linear chain polymer is involved and low density when the branched formed is used. The high density polyethylene has the greater crystallinity.

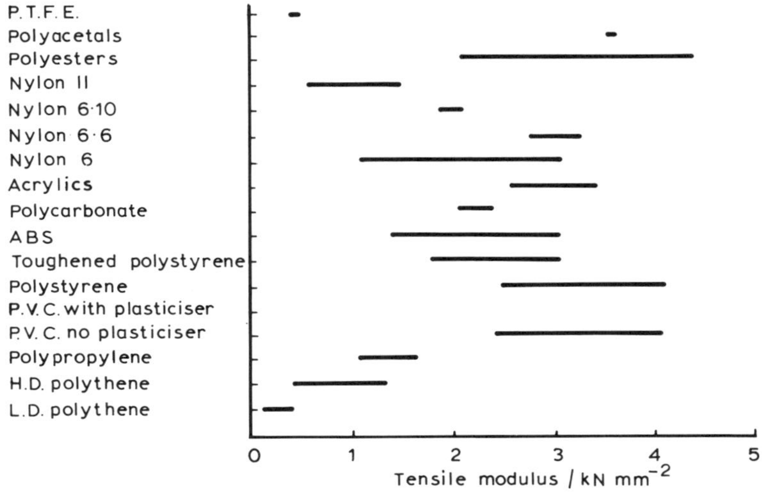

Figure 5.11 Comparison of tensile modulus values for thermoplastics

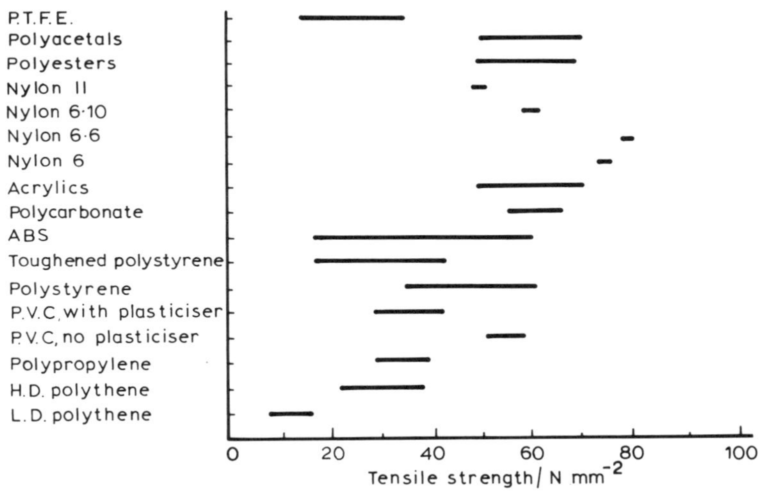

Figure 5.12 Comparison of tensile strength values for thermoplastics

Property	*Low density polyethylene*	*High density polyethylene*
Crystallinity	60%	95%
Density/10^3 kg m^{-3}	0.92	0.95
T_g/°C	−120	−120
T_m/°C	115	138
Tensile strength/N mm^{-2}	8 to 16	22 to 38
Tensile modulus/kN mm^{-2}	0.1 to 0.3	0.4 to 1.3
Elongation %	100 to 600	50 to 800

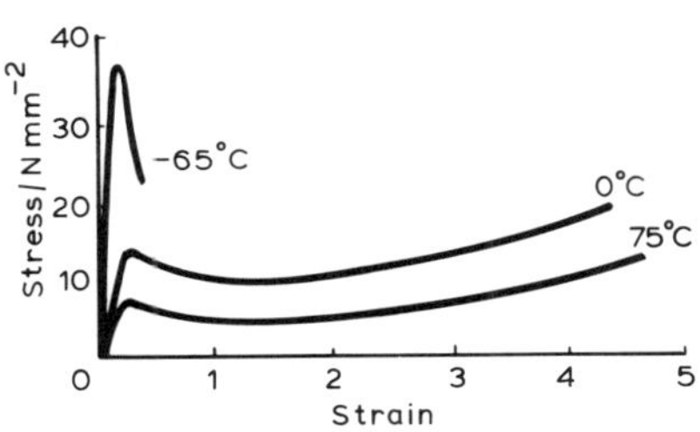

Figure 5.13 Stress–strain graph for low density polythene

Low- and high-density polymers may be blended to give plastics with properties intermediate between those quoted above. *Figure 5.13* shows the stress–strain graph for low-density polythene.

Low-density polyethylene is used mainly in the form of films and sheeting, e.g. polythene bags, 'squeeze' bottles, ball-point pen tubing, wire and cable insulation. High-density polythene is used for piping, toys, filaments for fabrics, household ware. Both forms have excellent chemical resistance, low moisture absorption and high electrical resistance. The additives commonly used with polyethylene

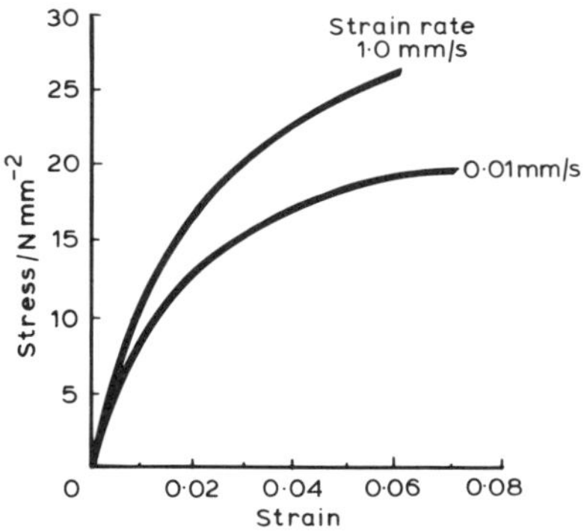

Figure 5.14 Stress-strain graph for polypropylene

are carbon black as a stabiliser, pigments to give coloured forms, glass fibres to give increased strength and butyl rubber to prevent inservice cracking.

Polypropylene is used mainly in its crystalline form, being a linear polymer with side groups regularly arranged along the chain. The presence of these side groups gives a more rigid and stronger polymer (*Figure 5.14*) than polyethylene in its linear form.

Property	*Polypropylene*
Crystallinity	60%
Density/10^3 kg m^{-3}	0.90
T_g/°C	−10
T_m/°C	176
Tensile strength/N mm^{-2}	30 to 40
Tensile modulus/kN mm^{-2}	1.1 to 1.6
Elongation %	50 to 600

Polypropylene is used for crates, containers, fans, car fascia panels, tops of washing machines, cabinets for radios and televisions, toys, chair shells.

Polyvinyl chloride (p.v.c.) is a linear chain polymer with bulky side groups (see *Figure 5.4*) which prevent crystalline regions occurring. The material is hard and rigid. It is often used with a plasticiser to give a more flexible plastic.

Property	*With no plasticiser*	*Low plasticiser content*
Crystallinity	0	0
Density/10^3 kg m^{-3}	1.4	1.3
T_g/°C	87	—
T_m/°C	—	—
Tensile strength/N mm^{-2}	52 to 58	28 to 42
Tensile modulus/kN mm^{-2}	2.4 to 4.1	—
Elongation %	2 to 40	200 to 250

The rigid form of p.v.c., i.e. the unplasticised form, is used for piping for waste and soil drainage systems, rainwater pipes, lighting fittings, curtain rails. Plasticised p.v.c. is used for the fabric of 'plastic' raincoats, bottles, shoe soles, garden hose piping, gaskets, inflatable toys.

Polystyrene is a linear chain polymer with bulky side groups which prevent crystalline regions occurring. Polystyrene with no additives is a brittle, transparent, material. A toughened form of polystyrene is produced by blending polystyrene with rubber particles.

Property	*No additives*	*Toughened*
Crystallinity	0	0
Density/10^3 kg m^{-3}	1.1	1.1
T_g/°C	100	—
T_m/°C	—	—
Tensile strength/N mm^{-2}	35 to 60	17 to 42
Tensile modulus/kN mm^{-2}	2.5 to 4.1	1.8 to 3.1
Elongation %	1 to 3	8 to 50

Polystyrene is an excellent electrical insulator and is widely used in electrical equipment. Other applications are packaging of cosmetics, toys, boxes. A widely used form of polystyrene is as expanded polystyrene, which is a rigid foam used widely for insulation and packaging. Toughened polystyrene is less brittle than ordinary polystyrene and finds use as cups in vending machines, casing for cameras, projectors, radios, television sets, vacuum cleaners.

Acrylonitrile-butadiene-styrene terpolymer (ABS) is produced by forming polymer chains with three different polymer materials, polystyrene, acrylonitrile and butadiene. It gives an amorphous material which is tough, stiff (see *Figure 1.14* and *Figure 5.18*) and abrasion resistant.

Property	*ABS*
Crystallinity	0
Density/10^3 kg m^{-3}	1.1
T_g/°C	71 to 112
T_m/°C	—
Tensile strength/N mm^{-2}	17 to 58
Tensile modulus/kN mm^{-2}	1.4 to 3.1
Elongation %	10 to 140

ABS is widely used as the casing for telephones, vacuum cleaners, hair driers, radios, televisions, typewriters, luggage, boat shells, food containers.

Polycarbonate is an amorphous thermoplastic having a linear chain with bulky side groups. It is tough, stiff and strong.

Property	*Polycarbonate*
Crystallinity	0
Density/10^3 kg m^{-3}	1.2
T_g/°C	150
T_m/°C	—
Tensile strength/N mm^{-2}	55 to 65
Tensile modulus/kN mm^{-2}	2.1 to 2.4
Elongation %	60 to 100

It is used for applications where the plastics required are resistant to impact abuse and, for plastics, relatively high temperatures. Typical applications are transparent street lamp covers, infant-feeding bottles, machine housings, safety helmets, housings for car lights, and tableware such as cups and saucers.

Acrylics are completely transparent thermoplastics having linear chains with bulky side groups and so giving an amorphous structure. They give a stiff, strong, material with outstanding weather resistance.

Property	*Acrylics*
Crystallinity	0
Density/10^3 kg m^{-3}	1.18
T_g/°C	0
T_m/°C	—
Tensile strength/N mm^{-2}	50 to 70
Tensile modulus/kN mm^{-2}	2.7 to 3.5
Elongation %	5 to 8

Because of its transparency it is used for windscreens, light fittings, canopies, lenses for car lights, signs and nameplates. Opaque acrylic sheet is used for the production of domestic baths, shower cabinets, basins and lavatory cisterns.

Polyamides, or nylons as they are better known, are linear polymers and give crystalline structures. There are a number of common polyamides: nylon 6, nylon 6.6, nylon 6.10 and nylon 11. The numbers refer to the numbers of carbon atoms in each of the reacting substances used to give the polymer. The full stops separating the two numbers are sometimes omitted, e.g. nylon 66 is nylon 6.6. The two most used nylons are nylon 6 (see *Figure 1.36*) and nylon 6.6. Nylon 6.6 has a higher melting point than nylon 6 and is also stronger and stiffer. Nylon 11 has a lower melting point than either nylon 6 or 6.6; it is also more flexible. Nylon 6.10 has properties intermediate between nylon 11 and nylon 6.6.

In general, nylon materials are strong, tough and have relatively high melting points. But they do tend to absorb moisture (see *Figure 5.17*). The effect of which is to reduce their tensile strength. Nylon 6.6 can absorb quite large amounts of moisture, nylon 11 however absorbs considerably less.

Property	*Nylon 6*	*Nylon 6.6*	*Nylon 6.10*	*Nylon 11*
Crystallinity	Can be varied from low to high percentages			
Density/10^3 kg m^{-3}	1.13	1.1	1.1	1.1
T_g/°C	50	57	50	
T_m/°C	225	265	228	
Tensile strength/N mm^{-2}	75	80	60	50
Tensile modulus/kN mm^{-2}	1.1 to 3.1	2.8 to 3.3	1.9 to 2.1	0.6 to 1.5
Elongation %	60 to 320	60 to 300	85 to 230	70 to 300

Nylons often contain additives, e.g. a stabiliser or flame retardant substance. Glass spheres or glass fibres are added to give improved strength and rigidity. Molybdenum disulphide is an additive to nylon 6 to give a material with very low frictional properties.

Nylon is used for the manufacture of fibres for clothing, gears, bearings, bushes, housings for domestic and power tools, electric plugs and sockets.

Polyesters are available in a thermoplastic form, usually *polyethylene terephthalate.* This is a linear chain polymer with side groups. It gives crystalline structures and is below its glass transition point at normal temperatures. If it is rapidly quenched from a melt to below its glass transition point an amorphous structure is produced, the molecular chains not having sufficient time to become orderly packed.

Property	*Polyethylene terephthalate*
Crystallinity	60%
Density/10^3 kg m^{-3}	1.3
T_g/°C	69
T_m/°C	267
Tensile strength/N mm^{-2}	50 to 70
Tensile modulus/kN mm^{-2}	2.1 to 4.4
Elongation %	60 to 100

The polyester has properties similar to nylon. It is used widely in fibre form for the production of clothes. Other uses are for electrical plugs and sockets, push-button switches, wire insulation, recording tapes, insulating tapes, gaskets.

Polyacetals, polyformaldehyde or polyoxymethylene, are linear polymers giving rise to crystalline structures. They are stiff, strong, polymers and maintain their properties at relatively high temperatures. They are however adversely affected by ultraviolet light and thus have to be used with a stabiliser.

Property	*Polyformaldehyde*	*Polyoxymethylene*
Crystallinity	High	High
Density/10^3 kg m^{-3}	1.41	1.41
T_g/°C	−73	−76
T_m/°C	180	180
Tensile strength/N mm^{-2}	50 to 70	70
Tensile modulus/kN mm^{-2}	3.6	3.6
Elongation %	15 to 75	15 to 75

Typical applications are pipe fittings, parts for water pumps and washing machines, car instrument housings, bearings, gears, hinges and window catches, seat belt buckles.

Polytetrafluoroethylene (P.T.F.E.) is a linear polymer like polyethylene, the only difference being that instead of hydrogen atoms there are fluorene atoms. It has a very high crystallinity as manufactured, about 90%, though this degree of crystallinity can be reduced during processing to about 50% if quench cooled, or 75% with slow cooling. Tough and flexible, p.t.f.e. can be used over a wide range of temperature, 250°C down to almost absolute zero, and still retain the very important property of not being attacked by any reagent or solvent. It also has a very low coefficient of friction. No known material can be used to bond it satisfactorily to other materials.

Property	*P.T.F.E.*
Crystallinity	90%
Density/10^3 kg m^{-3}	2.2
T_g/°C	−120
T_m/°C	327
Tensile strength/N mm^{-2}	14 to 35
Tensile modulus/kN mm^{-2}	0.4
Elongation %	200 to 600

P.T.F.E. is a relatively expensive material and is not processed as easily as other thermoplastics. It tends to be used where its special properties, i.e. not attacked chemically and very low coefficient of friction, are needed. Journal bearings with a p.t.f.e. surface can be used without lubrication because of the low coefficient of friction; they can even be used at temperatures up to about 250°C. Piping carrying corrosive chemicals at temperatures up to 250°C are made of p.t.f.e. Other applications for p.t.f.e. are gaskets, diaphragms, valves, O-rings, bellows, couplings, dry and self-lubricating bearings, coatings for frying pans and other cooking utensils (known as 'non-stick'), coverings for rollers handling sticky materials, linings for hoppers and chutes, and electrical insulating tape.

THERMOSETTING POLYMERS

Thermoplastic polymers soften if heated, and can be reshaped, the new shape being retained when the plastic cools. The process can be repeated. *Thermosetting polymers* cannot be softened and reshaped by heating. They are plastic in the initial stage of manufacture but once they have set they cannot be resoftened. The atoms in a thermosetting material form a three-dimensional structure of cross-linked chains (see *Figure 5.1c*). The bonds linking the chains are strong and not easily broken. Thus the chains cannot slide over one another but are essentially fixed in the positions they occupied when the polymer was solidifying during its formation.

Thermosetting polymers are stronger and stiffer than thermoplastics and generally they can be used at higher temperatures than thermoplastics. As they cannot be shaped after the initial reaction in which the polymer chains are produced, the processes by which thermosetting polymers can be shaped are limited to those where the product is formed from the raw polymer materials. No further processing is possible (other than possibly some machining) and this limits the processes available to essentially just *moulding*. A number of different moulding methods are used but essentially all involve the combining together of the chemicals in a mould so that the cross-linked chains are produced while the material is in the mould. The result is a thermosetting polymer shaped to the form dictated by the mould.

PROPERTIES OF COMMON THERMOSETS

Phenolics give highly cross-linked polymers. *Phenol formaldehyde* was the first synthetic plastic and is known as *Bakelite.* The polymer is opaque and initially light in colour. It does however darken with time and so is always mixed with dark pigments to give dark-coloured materials. It is supplied in the form of a moulding powder,

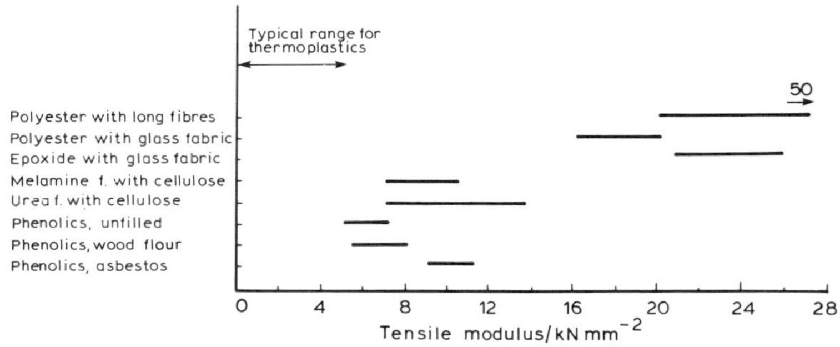

Figure 5.15 Comparison of tensile modulus values of thermosets

including the resin, fillers and other additives such as pigments. When this moulding powder is heated in a mould the cross-linked polymer chain structure is produced. The fillers account for some 50 to 80% of the total weight of the moulding powder. Wood flour, a very fine soft wood sawdust, when used as a filler increases the impact strength of the plastic, asbestos fibres improve the heat properties, and mica the electrical resistance.

Property	*Unfilled*	*Wood flour filler*	*Asbestos filler*
Density/10^3 kg m^{-3}	1.25 to 1.30	1.32 to 1.45	1.6 to 1.85
Tensile strength/N mm^{-2}	35 to 55	40 to 55	30 to 55
Tensile modulus/kN mm^{-2}	5.2 to 7.0	5.5 to 8.0	C.1 to 11.5
Elongation %	1 to 1.5	0.5 to 1	0.1 to 0.2
Max. service temp./°C	120	150	180

Phenol formaldehyde mouldings are used for electrical plugs and sockets, switches, door knobs and handles, camera bodies and ash trays. Composite materials (see Chapter 6) involving the phenolic resin being used with paper or an open weave fabric, e.g. a glass

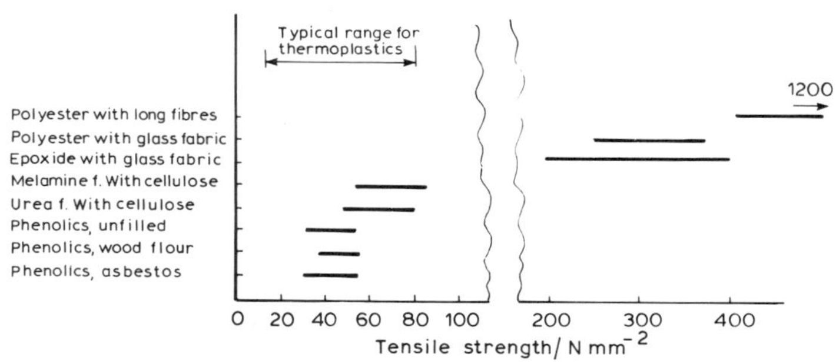

Figure 5.16 Comparison of tensile strength values of thermosets

fibre fabric, are used for gears, bearings, and electrical insulation parts.

Amino-formaldehyde materials, generally *urea formaldehyde* and *melamine formaldehyde,* give highly cross-linked polymers. Both are used as moulding powders, like the phenolics. Cellulose and wood flour are widely used as fillers. Hard, rigid, high-strength materials are produced.

Property	*Urea formaldehyde cellulose filler*	*Melamine formaldehyde cellulose filler*
Density/10^3 kg m^{-3}	1.5 to 1.6	1.5 to 1.6
Tensile strength/N mm^{-2}	50 to 80	55 to 85
Tensile modulus/kN mm^{-2}	7.0 to 13.5	7.0 to 10.5
Elongation %	0.5 to 1.0	0.5 to 1.0
Max. service temp./°C	80	95

Both materials are used for table ware, e.g. cups and saucers, knobs, handles, light fittings and toys. Composites with open weave fabrics are used as building panels and electrical equipment.

Epoxide materials are generally used in conjunction with glass, or other, fibres to give hard and strong composites. Epoxy resins are excellent adhesives giving very high adhesive strengths.

Property	*Plain weave glass fabric*
Percentage fabric	60 to 65%
Density/10^3 kg m^{-3}	1.8
Tensile strength/N mm^{-2}	200 to 420
Tensile modulus/kN mm^{-2}	21 to 25
Max. service temp./°C	200

The unfilled epoxide has a tensile strength of 35 to 80 N mm^{-2}, considerably less than that of the composite. The composite is used for boat hulls and table tops.

Polyesters can be produced as either thermosets or thermoplastics (see page 82). The thermoset form is mainly used with glass, or other, fibres to hard and strong composites. Such composites are used for boat hulls, architectural panels, car bodies, panels in aircraft, and stackable chairs. The properties of the composites will be found on page 99. They have a maximum service temperature of the order of 200°C.

ELASTOMERS

Elastomers are polymers which show very large strains when subject to stress and which will return to their original dimensions when the stress is removed. Elastomers are essentially amorphous polymers with a glass transition temperature below their service temperature. The polymer structure is that of linear chain molecules with some cross-linking between chains (*Figure 5.17*). Without the cross-linking, the molecular chains might slide past each other and give permanent deformations. The presence of these cross-links ensures that the material is elastic and returns to its original dimensions when the stress is removed. However if there are too many cross-links the material becomes inflexible.

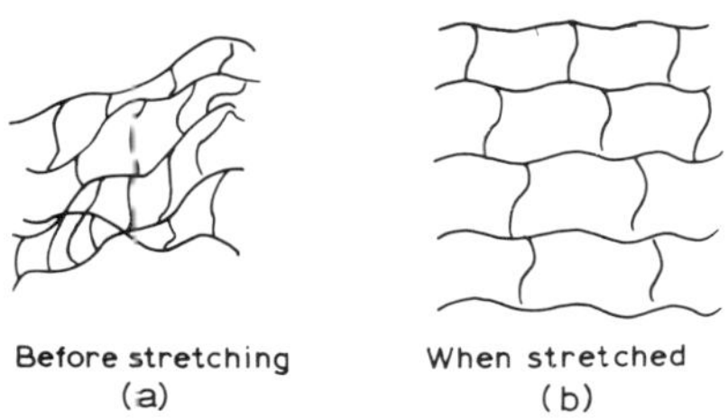

Figure 5.17 Stretching an elastomer which has cross links

PROPERTIES OF COMMON ELASTOMERS

Natural rubber is, in its crude form, just the sap from a particular tree. The addition of sulphur to the rubber produces cross-links, the amount of cross-linkage being determined by the amount of sulphur added. The process of producing cross-links is called *vulcanisation.* Anti-oxidants and plasticisers are also added to the rubber.

Property	*Natural rubber*
Tensile strength/N mm^{-2}	20
Elongation %	800
T_g/°C	−73
Service temp. range/°C	−50 to +80
Resistance to oils and greases	Poor
Resistance to water	Good

Butadiene styrene rubbers, commonly called *SBR* or *GR-S* or *Buna S* rubbers, are synthetic rubbers. They are cheaper than natural rubber. It is used in the manufacture of tyres, hosepipes, conveyor belts and cable insulation.

Property	*SBR*
Tensile strength/N mm^{-2}	24
Elongation %	600
T_g/°C	−58
Service temp. range/°C	−50 to +80
Resistance to oils and greases	Poor
Resistance to water	Good

Butyl rubbers, known as *isobutylene isoprene* or *GR-I,* have the important property of extreme impermeability to gases. They are thus widely used for the inner linings of tubeless tyres, steam hoses and diaphragms.

Property	*Butyl*
Tensile strength/N mm^{-2}	20
Elongation %	900
T_g/°C	−53
Service temp. range/°C	−50 to +100
Resistance to oils and greases	Poor
Resistance to water	Good

Nitrile rubbers, known as *butadiene acrylonitrile* or *Buna N,* are extremely resistant to organic liquids and used for such applications as hoses, gaskets, seals, tank linings, rollers, valves.

Property	*Butyl*
Tensile strength/N mm^{-2}	28
Elongation %	700
T_g/°C	−10
Service temp. range/°C	−50 to +125
Resistance to oils and greases	Excellent
Resistance to water	Good

Neoprenes, known as *polychloroprene,* has good resistance to oils and a variety of other chemicals, also having good weathering characteristics. It is used for oil and petrol hoses, gaskets, seals, diaphragms and chemical tank linings.

Property	*Neoprene*
Tensile strength/N mm^{-2}	25
Elongation %	1000
T_g/°C	−48
Service temp. range/°C	−50 to +100
Resistance to oils and greases	Good
Resistance to water	Fair

PROBLEMS

1. How is the structure of an amorphous polymer different from that of a crystalline polymer?

2. How does the form of the polymer molecular chain determine the degree of crystallinity possible with the polymer?

3. Explain how the crystallinity of a polymer affects its properties.

4. What is the glass transition temperature?

5. P.V.C. has a glass transition temperature of 87°C. How would its properties below this temperature differ from those above it?

6. A polypropylene article was designed for use at room temperature. What difference might be expected in its behaviour if used at about −15°C. The glass transition temperature for polypropylene is −10°C.

7. When a piece of polythene is pulled it starts necking at one point. Further pulling results in no further reduction in the cross-section of the material at the necked region but a spread of the necked region along the entire length of the material. Why doesn't the material just break at the initial necked section instead of the necking continuing?

8. Why are the properties of a cold drawn polymer different from those of the undrawn polymer?

9. Why are polyester fibres cold drawn before use?

10. Polypropylene is a crystalline polymer with a glass transition temperature of −10°C and a melting point of 176°C. What would be its normal maximum service temperature? At what temperatures would the polymer be hot formed?

11. Explain what is meant by internal and external plasticisation.

12. What is the effect of a plasticiser on the mechanical properties of a polymer?

13. How do the properties of high and low density polythene differ?

14. How do the properties of polystyrene with no additives and toughened polystyrene differ?

15. To what temperature would high density polythene have to be heated to be hot formed?

16. Which is stiffer at room temperature, p.v.c. with no plasticiser or polypropylene?

17. The casing of a telephone is made from ABS. How would the casing behave if someone left a burning match or cigarette against it?

18. What are the special properties of p.t.f.e. which render it useful despite its high price and processing problems?

19. How do the mechanical properties of thermosets differ, in general, from those of thermoplastics?

20. What is bakelite and what are its mechanical properties?

21. Describe how cups made of melamine formaldehyde with a cellulose filler might be expected to behave in service.

22. What is meant by vulcanisation?

23. Would Buna S or Buna N be the best rubber to use for a pipe carrying oil?

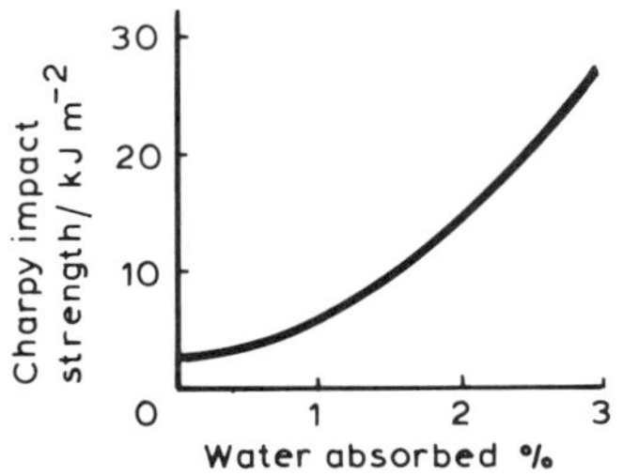

Figure 5.18 The effect of water absorption on the impact strength of nylon 6

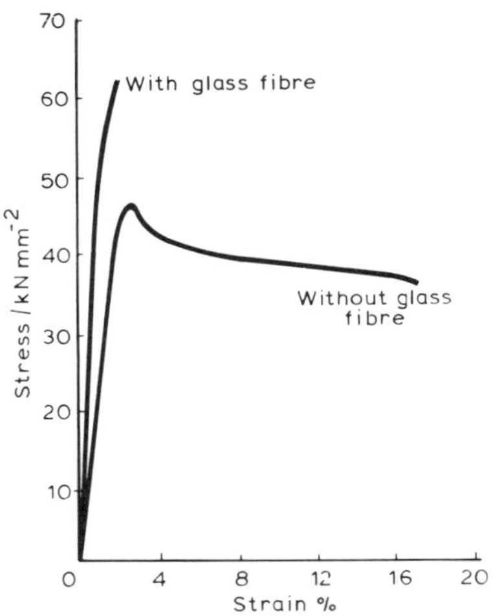

Figure 5.19 Stress–strain graphs for ABS plastics Novodur PHGV (reinforced) and PH-AT (un-reinforced). (Courtesy of Bayer UK Ltd)

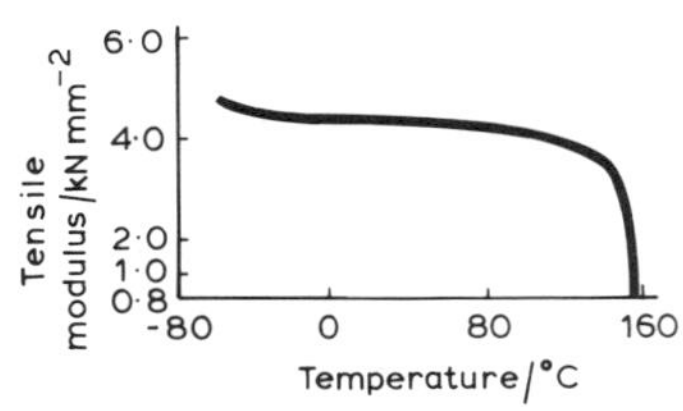

Figure 5.20 The effect of temperature on the tensile modulus of Makrolon 2800. (Courtesy of Bayer UK Ltd)

24. Explain how rubbers can be stretched to several times their length and return to the same initial length when released.

25. *Figure 5.18* shows the effect on the Charpy impact strength for nylon 6 of percentage of water absorbed. As the percentage of water absorbed increases, is the material becoming more or less brittle?

26. *Figure 5.19* shows the stress–strain graphs for two forms of ABS plastic, one containing 20% by weight glass fibre and the other without such fibres. Estimate from the graph (a) the tensile strength and (b) the tensile modulus for both forms.

27. Explain the significance of the following information from a manufacturer's data sheet for neoprene rubber. (Courtesy of Du Pont Ltd.)

Recovery from deformation

The recovery of a material after being held under load is important in the design of many rubber products. Neoprene exhibits a relatively low degree of permanent deformation from compression (compression set). For example, specimens of a standard compound, compressed to 75% of their original thickness and then released behaved as follows:

Held 22 hours at 70°C–15% compression set.

Held 70 hours at 100°C–35% compression set.

Permanent deformation due to elongation (permanent set) will be approximately 5% for most neoprene products.

28. *Figure 5.20* shows how the tensile modulus of a polycarbonate plastic Makrolon 2800 changes with temperature. (Courtesy of Bayer Ltd.) How is the behaviour of the material under stress changing as the temperature increases?

29. Which polymers would be suitable for the following applications? (a) An ash tray. (b) A garden hose pipe. (c) A steam hose. (d) Insulation for electric wires. (e) The transparent top of an electrical meter. (f) A plastic raincoat. (g) A toothbrush. (h) A camera body.

6 Composites

Objectives: At the end of this chapter you should be able to:
Give examples of common composites met in engineering.
Explain how the mechanical properties of composites are related to fibre direction and properties of the matrix and fibre.

EXAMPLES OF COMPOSITES

The team *composite* is used for a material composed of two different materials bonded together with one serving as the matrix surrounding fibres or particles of the other. A common example of a composite is reinforced concrete. This has steel rods embedded in the concrete (*Figure 6.1*). The composite enables loads to be carried that otherwise

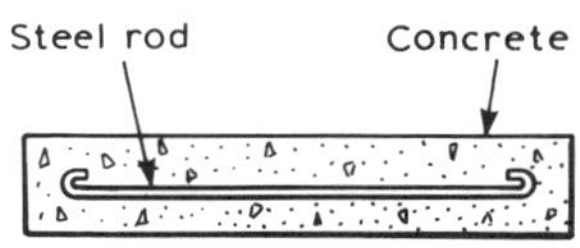

Figure 6.1 Reinforced concrete, steel rods in a matrix

Figure 6.2 Concrete aggregate in a matrix

could not have been carried by the concrete alone. Concrete itself is a composite, without the presence of steel reinforcement. It is made by mixing cement, sand, aggregate and water. Stone chips or gravel are often used as the aggregate. The resulting concrete consists of the aggregate in a matrix (*Figure 6.2*).

There are many examples of composite materials encountered in everyday components. Many plastics are glass fibre or glass particle reinforced. Vehicle tyres are rubber reinforced with woven cords. Wood is a natural composite material with tubes of cellulose bonded by a natural plastic called lignin (*Figure 6.3*). Cermets, widely used for cutting tool tips, are composites involving ceramic particles in a metal matrix.

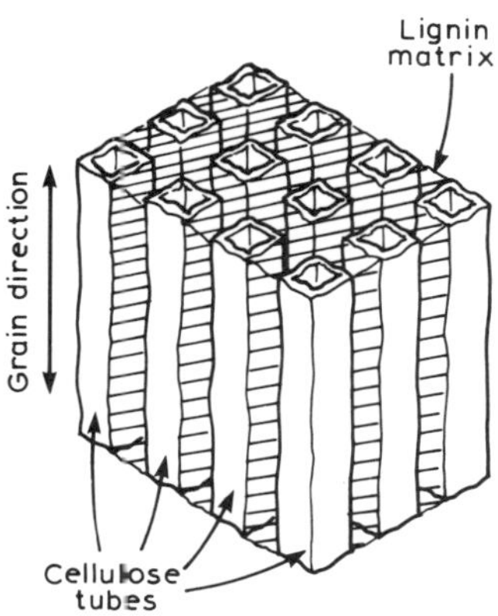

Figure 6 3 Wood, cellulose fibres in a lignin matrix

CONTINUOUS FIBRES IN A MATRIX

Consider a composite rod made up of continuous fibres, all parallel to the rod axis, in a matrix (*Figure 6.4*). These could be glass fibres in a plastic or steel reinforcement rods in concrete. Each element in the composite has a share of the applied force, thus

$$\text{Total force} = \text{force on fibres} + \text{force on matrix}$$

But the stress on the fibres is equal to force on them divided by their cross-sectional area. Similarly the stress on the matrix is equal to the force on the matrix divided by its area. Hence

$$\text{Total force} = \text{stress on fibres} \times \text{area of fibres} + \text{stress on matrix} \times \text{area of matrix}$$

Dividing both sides of the equation by the total area of the composite gives

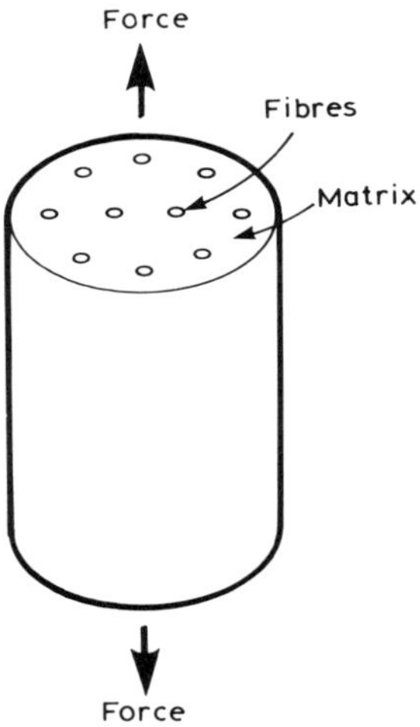

Figure 6.4 Continuous fibres in a matrix

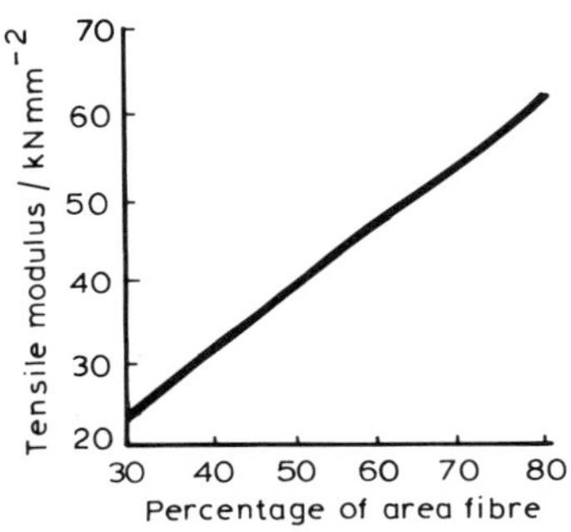

Figure 6.5 The effect of the percentage of cross-sectional area of glass fibre on the tensile modulus of a glass fibre–polyester composite

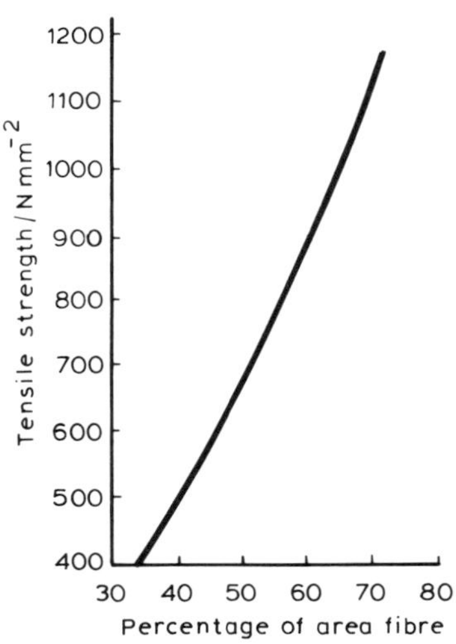

Figure 6.6 The effect of the percentage of cross-sectional area of glass fibre on the tensile strength of a glass fibre–polyester composite

$$\frac{\text{Total force}}{\text{total area}} = \text{stress on fibres} \times \frac{\text{area of fibres}}{\text{total area}} + \text{stress on matrix} \times \frac{\text{area of matrix}}{\text{total area}}$$

The fraction of the cross-sectional area that is fibre is given by the area of the fibres divided by the total area, similarly the fraction of the cross-section that is matrix is the area of the matrix divided by the total area. The total area divided by the total force is the stress applied to the composite. Thus

Stress on composite = stress on fibres × area fraction fibres + stress on matrix × area fraction matrix

If the fibres are firmly bonded to the matrix then the elongation of contraction of the fibres and matrix must be the same and equal to that of the composite as a whole. Thus

Strain on composite = strain on fibres = strain on matrix

Dividing both sides of the stress equation by the strain gives an equation in terms of the tensile modulii (stress/strain = tensile modulus). Thus

Modulus of composite = modulus of fibres × area fraction fibres + modulus of matrix × area fraction matrix

Suppose we have glass fibres with a tensile modulus of 76 kN mm^{-2} in a matrix of polyester having a tensile modulus of 3 kN mm^{-2}. Then, if the fibres occupy 60% of the cross-sectional area, the tensile modulus of the composite will be given by

$$\text{Modulus of composite} = 76 \times 0.6 + 3 \times 0.4$$
$$= 46.8 \text{ kN mm}^{-2}$$

The composite has a tensile modulus considerably greater than that of the polyester.

Not only has the composite a higher tensile modulus but also a higher strength than that of the matrix material. Thus for the 60% glass fibres in polyester the tensile strength may be about 800 N mm^{-2} for the composite when the matrix has a tensile strength of only about 50 N mm^{-2}.

For glass fibre–polyester composites with long fibres all in the same direction and all parallel to the axis of the composite along which the force is applied, *Figure 6.5* shows how the tensile modulus depends on the percentage of the area occupied by the fibres and *Figure 6.6* how the tensile strength is affected by this factor. Such composites usually have between 40% and 80% of the cross-sectional area as glass fibre. The following is the data for the glass fibres and polyester when separate.

	Tensile modulus /kN mm^{-2}	*Tensile strength* /N mm^{-2}
Polyester	2 to 4	20 to 70
E-glass fibres	76	1200 to 1800

Example

A column of reinforced concrete has steel reinforcing rods running through the entire length of the column and parallel to the column axis. If the concrete has a modulus of elasticity of 20 kN mm^{-2} and the steel 210 kN mm^{-2}, and the steel rods occupy 10% of the cross-sectional area of the column, what is the modulus of elasticity for the composite?

$$\text{Modulus of composite} = 210 \times 0.1 + 20 \times 0.9 = 39 \text{ kN mm}^{-2}$$

Example
Carbon fibres with a tensile modulus of 400 kN mm^{-2} are used to reinforce aluminium with a tensile modulus of 70 kN mm^{-2}. If the fibres are long and parallel to the axis along which the load is applied, and occupy 50% of the cross-sectional area of the composite, what is the tensile modulus of the composite?

$$\text{Modulus of composite} = 400 \times 0.5 + 70 \times 0.5 = 235 \text{ kN mm}^{-2}.$$

REINFORCED PLASTICS

Reinforced plastics consist of a stiff, strong, material combined with the plastic. Glass fibres are probably the most used additive. The fibres may be long lengths, running through the length of the composite, or discontinuous short lengths randomly orientated within the composite. Another form of composite uses glass fibre mats or cloth in the plastic. The effect of the additives is to increase both the tensile strength and the tensile modulus of the plastic, the amount of change depending on both the form the additive takes and the amount of it. The continuous fibres give the highest tensile modulus and tensile strength composite but with a high directionality of properties. The strength along the direction of the fibres could be perhaps 800 N mm^{-2} while that at right-angles to the fibre direction may be as low as 30 N mm^{-2}, i.e. just about the strength of the plastic alone. Random orientated short fibres do not lead to this directionality of properties but do not give such high strength and tensile modulus. The composites with glass fibre mats or cloth tend to give tensile strength and modulus values intermediate between those of the continuous and short length fibres. The following are examples of the strength and modulus values obtained with reinforced polyester.

Material	*Percentage weight of glass*	*Tensile modulus* /kN mm^{-2}	*Tensile strength* /N mm^{-2}
Polyester	0	2 to 4	20 to 70
With short fibres	10 to 45	5 to 14	40 to 180
With plain weave cloth	45 to 65	10 to 20	250 to 350
With long fibres	50 to 80	20 to 50	400 to 1200

DISPERSION STRENGTHENED METALS

The strength of a metal can be increased by small particles dispersed throughout it. Thus solution treatment followed by precipitation hardening for an aluminium-copper alloy can lead to a fine dispersion of an aluminium–copper compound throughout the alloy. The result is a higher tensile strength material.

Aluminium alloy	*Condition*	*Tensile strength* /N mm^{-2}
4.0% Cu, 0.8% Mg,	Annealed	190
0.5% Si, 0.7% Mn.	Solution treated, precipitation hardened	430

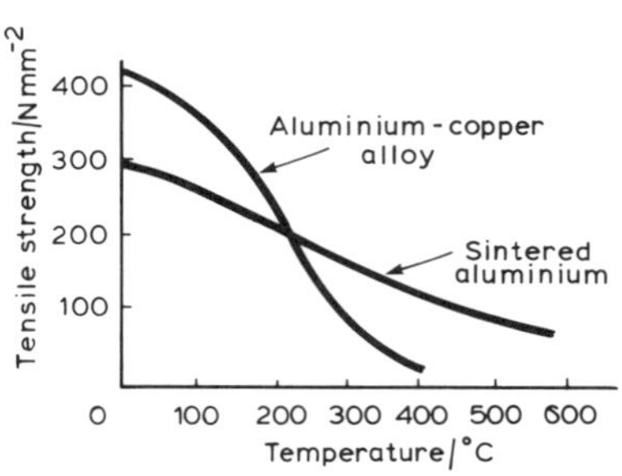

Figure 6.7 The effect of temperature on the tensile strength of an aluminium-copper alloy and sintered aluminium

Another way of introducing a dispersion of small particles throughout a metal involves *sintering*. This process involves compacting a powdered metal powder in a die and then heating it to a temperature high enough to knit together the particles in the powder. If this is done with aluminium the result is a fine dispersion of aluminium oxide (about 10%) throughout an aluminium matrix. The aluminium oxide occurs because aluminium in the presence of oxygen is coated with aluminium oxide. When the aluminium powder is compacted, much of the surface oxide film becomes separated from the aluminium and becomes a fine powder dispersed throughout the metal. The aluminium oxide powder, a ceramic, dispersed throughout the aluminium matrix gives a stronger material than that which would have been given by the aluminium alone. At room temperature the tensile strength of the sintered aluminium powder is about 400 N mm^{-2}, compared with that of about 90 N mm^{-2} for commercial, annealed, aluminium. The sintered aluminium has an advantage over the precipitation hardened aluminium alloy in that it retains its strength better at high temperatures (*Figure 6.7*). This is because, at the higher temperatures, the precipitate particles tend to coalesce or go into solution in the metal.

Other dispersion-strengthened metals involving ceramics, have been developed. Thorium oxide particles (about 7%) in nickel give a composite which has good strength properties at high temperatures.

CERMETS

Cermets are composites involving ceramics in a matrix of metal. The ceramics used have high strengths, high values of tensile modulus, high hardness, but are by themselves brittle substances. By comparison, the metals are weaker and less stiff, but they are ductile. By combining the ceramic with the metal, up to 80% ceramic, a composite can be produced which is strong, hard and tough. The composite has the ceramic in particle form in the metal matrix.

Typical ceramics used in cermets are carbides (tungsten, titanium, silicon, molybdenum), borides (chromium, titanium, molybdenum) and oxides (aluminium, chromium, magnesium), the metals being cobalt, chromium, nickel, iron or tungsten. A typical cermet used for cutting tool bits involves tungsten carbide in a matrix of cobalt.

LAMINATED MATERIALS

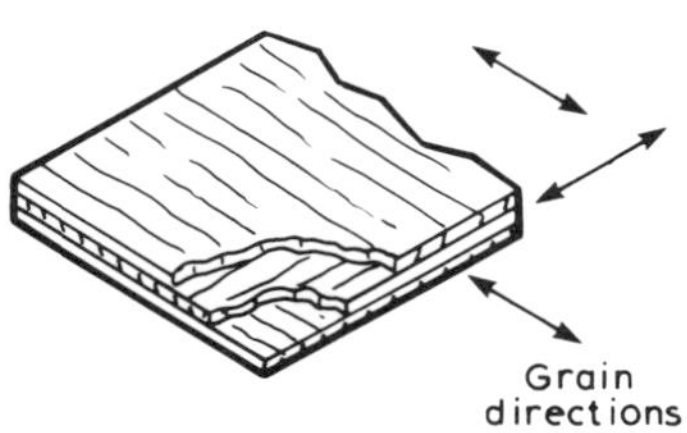

Figure 6.8 Plywood

Plywood is an example of a laminated material. It is made by gluing together thin sheets of wood with their grain directions at right angles to each other (*Figure 6.8*). The grain directions are the directions of the cellulose fibres in the wood and thus the resulting structure, the plywood, has fibres in mutually perpendicular directions. Thus whereas the thin sheet had properties that were directional the resulting laminate has no such directionality.

The term *laminated wood* is generally used to describe the product obtained by sticking together thin sheets of wood but with the grain of each layer parallel to the grain of the others. Large wooden arches and beams in modern buildings are likely to be laminated rather than a solid piece of wood. By carefully choosing the wood used to build up the beam a better quality beam can be produced than would otherwise be produced by nature.

It is not only wood that is laminated, metals are too. The *cladding*

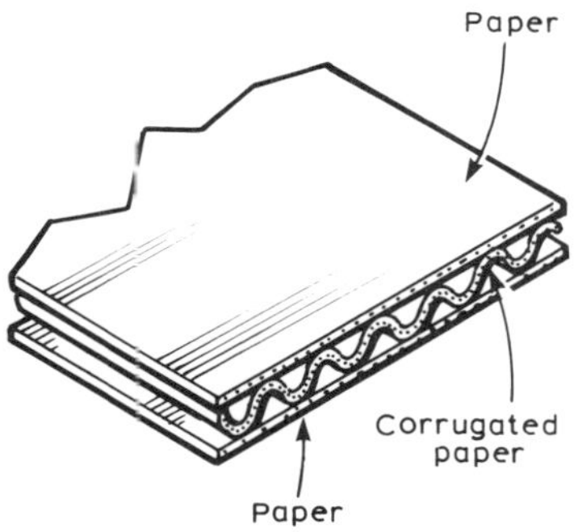

Figure 6.9 Corrugated cardboard

of aluminium-copper alloy with aluminium to give a material with a better corrosion resistance than that of the alloy alone is an obvious example. Galvanised steel can be considered another example, a layer of zinc on the steel in order to give better corrosion resistance. Steel for use in food containers is often plated with tin to improve corrosion resistance. Many of the metals clad or plated with other metals are in this form for improved corrosion resistance.

Corrugated cardboard is another form of laminated structure (*Figure 6.9*), consisting of paper corrugations sandwiched between layers of paper. The resulting structure is much stiffer, in the direction parallel to the corrugations, than the paper alone. A similar type of material is produced with metals, a metal *honeycomb structure* sandwiched between thin sheets of metal. Such a structure has good stiffness and is very light. Aluminium is often used for both the honeycomb and the sheets.

PROBLEMS

1. Explain the term composite.

2. Give examples of composites involving (a) plastics and (b) metals.

3. Describe how the mechanical properties of a fibre composite depend on the form of the fibres and their orientation.

4. Calculate the tensile modulus of a composite consisting of 45% by volume of long glass fibres, tensile modulus 76 kN mm^{-2}, in a polyester matrix, tensile modulus 4 kN mm^{-2}. In what direction does your answer give the modulus?

5. In place of the glass fibres referred to in Question 4, carbon fibres are used. What would be the tensile modulus of the composite if the carbon fibres had a tensile modulus of 400 kN mm^{-2}?

6. What is a cermet?

7. Explain how plywood gets its stiffness.

8. Explain how corrugated cardboard gets its stiffness.

9. Long boron fibres, tensile modulus 340 kN mm^{-2}, are used to make a composite with aluminium as the matrix, tensile modulus 70 kN mm^{-2}. What would be the tensile modulus of the composite in the direction of the fibres if the fibres constitute 50 per cent of the volume of the composite?

10. Explain what is meant by the term 'dispersion strengthened metals' and give an example of one.

Butterworths Checkbooks

Checkbook General Editors

for **Mathematics, Sciences, Electrical, Electronic, Telecommunications, Mechanical, Production, Marine and Motor Vehicle Engineering**

John O Bird and Anthony J C May of Highbury College of Technology, Portsmouth.

for **Building Construction, Civil Engineering, Surveying and Architecture**

Colin R Bassett, lately of Guildford County College of Technology.

Mathematics

•Mathematics 1 Checkbook

J O Bird and A J C May

1981 168 pages approx 186 × 123 mm
0 408 00609 9 Limp Illustrated
0 408 00632 3 Cased

•Mathematics 2 Checkbook

J O Bird and A J C May

1981 272 pages approx 186 × 123 mm
0 408 00610 2 Limp Illustrated
0 408 00633 1 Cased

•Mathematics 3 Checkbook

J O Bird and A J C May

1981 196 pages approx 186 × 123 mm
0 408 00611 0 Limp Illustrated
0 408 00634 X Cased

•Mathematics 4 Checkbook

J O Bird and A J C May

1981 240 pages approx 186 × 123 mm
0 408 00612 9 Limp Illustrated
0 408 00660 9 Cased

•Engineering Mathematics and Science 3 Checkbook

J O Bird, A J C May and D Ayling

1982 128 pages approx 186 × 123 mm
0 408 00625 0 Limp Illustrated

Sciences

•Chemistry 2 Checkbook

P J Chivers

1981 144 pages approx 186 × 123 mm
0 408 00622 6 Limp Illustrated
0 408 00637 4 Cased

•Chemistry 3 Checkbook

P J Chivers

1982 144 pages approx 186 × 123 mm
0 408 00658 7 Limp Illustrated
0 408 00662 5 Cased

•Building Science and Materials 2 Checkbook
and
Environmental Science 3/4 Checkbook
(see under Building Construction, etc. below)

•Electrical Science 3 Checkbook
(see under Electrical, Electronics, etc. below)

•Engineering Mathematics and Science 3 Checkbook
(see under Mathematics above)

•Mechanical Science 3 Checkbook
(see under Mechanical, Production, etc. below)

Building Construction, Civil Engineering, Surveying and Architecture

---

•Construction Drawing 1 Checkbook

J Greening and A Bowers

1982 144 pages approx 186 × 123 mm
0 408 00646 3 Limp Illustrated

•Construction Technology 1 Checkbook

R Chudley

1981 160 pages approx 186 × 123 mm
0 408 00602 1 Limp Illustrated
0 408 00642 0 Cased

•Construction Technology 2 Checkbook

R Chudley

1981 160 pages approx 186 × 123 mm
0 408 00603 X Limp Illustrated
0 408 00664 1 Cased

•Building Science and Materials 2 Checkbook

M D W Pritchard

1981 116 pages approx 186 × 123 mm
0 408 00607 2 Limp Illustrated
0 408 00640 4 Cased

•Environmental Science 3/4 Checkbook

M D W Pritchard

1982 144 pages approx 186 × 123 mm
0 408 00608 0 Limp Illustrated
0 408 00663 3 Cased

•Geotechnics 4 Checkbook

R Whitlow

1982 144 pages approx 186 × 123 mm
0 408 00631 5 Limp Illustrated

•Building Law 4 Checkbook

A Galbraith

1982 112 pages approx 186 × 123 mm
0 408 00588 1 Limp Illustrated

•Economics for the Construction Industry 4 Checkbook

V J Seddon and G B Atkinson

1982 144 pages approx 186 × 123 mm
0 408 00655 2 Limp Illustrated

•Building Services and Equipment 4 Checkbook

F Hall

1981 144 pages approx 186 × 123 mm
0 408 00613 7 Limp Illustrated
0 408 00641 2 Cased

•Building Services and Equipment 5 Checkbook

F Hall

1981 144 pages approx 186 × 123 mm
0 408 00614 5 Limp Illustrated
0 408 00651 X Cased

Electrical, Electronic and Telecommunications Engineering

•Microelectronic Systems 1 Checkbook

R Vears

1981 84 pages approx 186 × 123 mm
0 408 00552 1 Limp Illustrated
0 408 00638 2 Cased

•Microelectronic Systems 2 Checkbook

R Vears

1982 160 pages approx 186 × 123 mm
0 408 00659 5 Limp Illustrated

•Electrical and Electronic Principles 2 Checkbook

J O Bird and A J C May

1981 168 pages approx 186 × 123 mm
0 408 00600 5 Limp Illustrated
0 408 00635 8 Cased

•Electrical and Electronic Applications 2 Checkbook

D W Tyler

1982 144 pages approx 186 × 123 mm
0 408 00616 1 Limp Illustrated
0 408 00661 7 Cased

•Electrical Principles 3 Checkbook

J O Bird and A J C May

1981 160 pages approx 186 × 123 mm
0 408 00601 3 Limp Illustrated
0 408 00636 6 Cased

•Electrical Science 3 Checkbook

J O Bird, A J C May and J R Penketh

1982 128 pages approx 186 × 123 mm
0 408 00626 9 Limp Illustrated
0 408 00657 9 Cased

•Electronics 2 Checkbook

S A Knight

1981 144 pages approx 186 × 123 mm
0 408 00615 3 Limp Illustrated
0 408 00639 0 Cased

•Electronics 3 Checkbook

S A Knight

1982 144 pages approx 186 × 123 mm
0 408 00623 4 Limp Illustrated

Mechanical, Production, Marine and Motor Vehicle Engineering

•Engineering Drawing 1 Checkbook

L N Jeary

1982 160 pages approx 186 × 123 mm
0 408 00647 1 Limp Illustrated

•Workshop Processes and Materials 1 Checkbook

R L Timings and T E Savage

1982 144 pages approx 186 × 123 mm
0 408 00621 8 Limp Illustrated

•Mechanical Science 3 Checkbook

D Ayling

1982 144 pages approx 186 × 123 mm
0 408 00649 8 Limp Illustrated
0 408 00665 X Cased